AF556183

Development of Fishermen Community through Technology

Development of Fishermen Community through Technology

George Hadwin

Development of Fishermen Community through Technology

ISBN 978-93-5111-929-6

Published in 2016 in India by

RANDOM PUBLICATIONS

4376-A/4B, Gali Murari Lal, Ansari Road
New Delhi-110 002
Phone : +9111-43580356, 011-23289044, 011-43142548
e-mail: sales@randompublications.com,
info@randompublications.com, randomexports@gmail.com

Type Setting by : Friends Media, Delhi-110089
Digitally Printed at : Replika Press Pvt. Ltd.

Preface

A fisherman or fisher is someone who captures fish and other animals from a body of water, or gathers shellfish. Worldwide, there are about 38 million commercial and subsistence fishermen and fish farmers. The term can also be applied torecreational fishermen and may be used to describe both men and women. Fishing has existed as a means of obtaining food since the Mesolithic period.

Fishermen use a variety of equipment to catch fish. This can include nets of various sizes, fishing lines and traps. Many also operate machinery designed to hoist captured fish onto a boat. Often, this job involves maintaining machinery in good working order, troubleshooting and repairing equipment during fishing trips and keeping the fishing vessel clean. In addition, some fishing jobs involve diving into a body of water and catching fish with spears, while others require the use of rakes to collect kelp and other types of water-based vegetation.

Fishing and the fisherman have also influenced Ancient Egyptian religion; mullets were worshiped as a sign of the arriving flood season. Bastet was often manifested in the form of a catfish. In ancient Egyptian literature, the method that Amun used to create the world is associated with the tilapia's method of mouth-brooding.

A commercial fisherman, also known as a fisher, uses equipment including nets, fishing rods and traps, to catch fish and other marine life that will be consumed by humans or used as animal feed or bait. Some work on large boats in deep water. Those vessels have large crews. Other fishermen work in shallow water on small boats with few other people.

– Author

Contents

1

Introduction

FISHERMAN

A fisherman or fisher is someone who captures fish and other animals from a body of water, or gathers shellfish. Worldwide, there are about 38 million commercial and subsistence fishermen and fish farmers. The term can also be applied torecreational fishermen and may be used to describe both men and women. Fishing has existed as a means of obtaining food since the Mesolithic period.

HISTORY

Fishing has existed as a means of obtaining food since the Mesolithic period. During the time of the Ancient Egyptians, fishermen provided the majority of food for Egyptians. Fishing had become a major means of survival as well as a business venture.

Fishing and the fisherman have also influenced Ancient Egyptian religion; mullets were worshiped as a sign of the arriving flood season. Bastet was often manifested in the form of a catfish. In ancient Egyptian literature, the method that Amun used to create the world is associated with the tilapia's method of mouth-brooding.

COMMERCIAL FISHERMEN

According to the FAO, there were about 39 million fishers in countries producing more than 200, 000 tonnes in 2012, which is nearly 140% the number in 1995. The total fishery production of 66 million tonnes equated to an average productivity of 3. 5 tonnes per person.

Most of this growth took place in Asian countries, where four-fifths of world fishers and fish farmers dwell. Most fishermen are men involved in offshore and deep-sea fisheries. Women fish in some regions inshore from small boats or collectshellfish and seaweed. In many artisanal fishing communities, women are responsible for making and repairing nets, post-harvestprocessing and marketing.

RECREATIONAL FISHERMEN

Recreational fishing is fishing for pleasure or competition. It can be contrasted with commercial fishing, which is fishing for economic profit, or subsistence fishing, which is fishing for survival. The most common form of recreational fishing is done with a rod, reel, line, hooks and any one of a wide range of baits. Lures are frequently used in place of bait. Some people make handmade lures, including plastic lures and artificial flies.

The practice of catching or attempting to catch fish with a hook is called angling. When angling, it is sometimes expected or required that the fish be caught and released. Big-game fishing is fishing from boats to catch large open-water species such as tuna, sharks and marlin. Noodling and trout tickling are also recreational activities.

FISHING COMMUNITIES

For some communities, fishing provides not only a source of food and work but also community and cultural identity. In the New Testament, Jesus is reported to have said to his disciples: Follow me, and I will make you fishers of men.

SAFETY ISSUES

The fishing industry is hazardous for fishermen. Between 1992 and 1999, US commercial fishing vessels averaged 78 deaths per year.

The main contributors to fatalities are:

- Inadequate preparation for emergencies
- Poor vessel maintenance and inadequate safety equipment
- Lack of awareness of or ignoring stability issues.

Many fishermen, while accepting that fishing is dangerous, staunchly defend their independence. Many proposed laws and additional regulation to increase safety have been defeated because fishermen oppose them.

Alaska's commercial fishermen work in one of the world's harshest environments. Many of the hardships they endure include isolated fishing grounds, high winds, seasonal darkness, very cold water, icing, and short fishing seasons, where very long work days are the norm. Fatigue, physical stress, and financial pressures face most Alaska fishermen through their careers. The hazardous work conditions faced by fishermen have a strong impact on their safety.

Out of 948 work-related deaths that took place in Alaska during 1990-2006, one-third (311) occurred to fishermen. This is equivalent to an estimated annual fatality rate of 128/100, 000 workers/year.

This fatality rate is 26 times that of the overall U. S. work-related fatality rate of approximately 5/100, 000 workers/year for the same time period. While the work-related fatality rate for commercial fishermen in Alaska is still very high, it does appear to be decreasing: since 1990, there has been a 51 percent decline in the annual fatality rate. The successes in commercial fishing are due in part to the U. S. Coast Guard implementing new safety requirements in the early 1990s. These safety requirements contributed to 96 percent of the commercial fishermen surviving vessel sinkings/capsizings in 2004, whereas in 1991, only 73 percent survived.

While the number of occupational deaths in commercial fishermen in Alaska has been reduced, there is a continuing pattern of losing 20 to 40 vessels every year. There are still about 100 fishermen who must be rescued each year from cold Alaska waters.

Successful rescue is still dependent on the expertly trained personnel of the US Coast Guard Search and Rescue operations, and such efforts can be hindered by the harshness of seas and the weather. Furthermore, the people involved in Search and Rescue operations are themselves at considerable risk for injury or death during these rescue attempts.

DUTIES OF FISHERMEN?

Fig. Fishermen often use large nets to capture many fish at once.

According to the Bureau of Labour Statistics, you can likely obtain a fishing job without a high school diploma and earn median pay of about $25, 590 a year, based on 2010 wage statistics. Many fishing positions are seasonal, allowing you to pursue additional employment opportunities at other times of the year. This job might require you to spend weeks at a time away from home, however.

BASIC DUTIES

Fishermen use a variety of equipment to catch fish. This can include nets of various sizes, fishing lines and traps. Many also operate machinery designed to hoist captured fish onto a boat. Often, this job involves maintaining machinery in good working order, troubleshooting and repairing equipment during fishing trips and keeping the fishing vessel clean.

In addition, some fishing jobs involve diving into a body of water and catching fish with spears, while others require the use of rakes to collect kelp and other types of water-based vegetation.

HANDLING FISH

After catching fish, fishermen usually measure fish and release any that don't meet their jurisdiction's or captain's size requirements. Likewise, they examine fish for defects or signs of ill health, discarding any that don't meet expectations.

They must also sort their catches, clean the fish and ensure that they don't spoil by storing them in refrigerated or ice-filled holds. Once the ship returns to port, fishermen typically unload the fish as well.

OPERATING A FISHING VESSEL

Higher-ranking fishermen sometimes act as captains of a vessel, operating the boat and hiring, firing and training crew members. An individual with this job must supervise the crew and decide which types of fish to catch and where to go catch them.

He also oversees the maintenance of the boat and the fishing equipment, ensuring that it stays in good shape. A captain also navigates the fishing vessel.

PREPARATION

You can get a job as a fisherman without any training, as many positions offer on-the-job training. You can improve your chances of obtaining a job by enrolling in a fishery programme, which can include instruction in fishery technology, navigation and seamanship, at a vocational school or a community college. According to the U. S. Department of Labour, these programmes last for about two years. Choosing a programme that provides hands-on practice opportunities might give you a leg up on other candidates.

If you want to captain or operate a commercial fishing vessel that weighs more than 200 tons, you'll have to complete a Coast Guard approved training course and obtain a commercial fishing license. In addition, your state will likely require you to obtain a commercial fishing permit.

FISHING

Fishing is the activity of trying to catch fish. Fishing sometimes takes place in the wild. Techniques for catching fish includehand gathering, spearing, netting, angling and trapping. The term fishing may be applied to catching other aquatic animals such as molluscs, cephalopods, crustaceans, andechinoderms.

The term is not normally applied to catching farmed fish, or to aquatic mammals, such as whales, where the term whaling is more appropriate.

Fig. Stilts fishermen, Sri Lanka.

Fig. Fishing with nets, Mexico.

According to FAO statistics, the total number of commercial fishermen and fish farmers is estimated to be 38 million. Fisheries and aquaculture provide direct and indirect employment to over 500 million people in developing countries. In 2005, the worldwide per capita consumption of fish captured from wild fisheries was 14.4 kilograms, with an additional 7.4 kilograms harvested from fish farms. In addition to providing food, modern fishing is also a recreational pastime.

HISTORY

Fishing is an ancient practice that dates back to at least the beginning of the Paleolithic period about 40,000 years ago. Isotopic analysis of the skeletal remains of Tianyuan man, a 40,000-year-old modern human from eastern Asia, has shown that he regularly consumed freshwater fish. Archaeology features such as shell middens, discarded fish bones, and cave paintings show that sea foods were important for survival and consumed in significant quantities. During this period, most people lived a hunter-gatherer lifestyle and were, of necessity, constantly on the move. However, where there are early examples of permanent settlements (though not necessarily permanently occupied) such as those at Lepenski Vir, they are almost always associated with fishing as a major source of food.

MODERN TRAWLING

The British dogger was an early type of sailing trawler from the 17th century, but the modern fishing trawler was developed in the 19th century, at the English fishing port of Brixham. By the early 19th century, the fishermen at Brixham needed to expand their fishing area further than ever before due to the ongoing depletion of stocks that was occurring in the overfished waters of South Devon. The Brixham trawler that evolved there was of a sleek build and had a tall gaff rig, which gave the vessel sufficient speed to make long distance trips out to the fishing grounds in the ocean.

They were also sufficiently robust to be able to tow large trawls in deep water. The great trawling fleet that built up at Brixham, earned the village the title of 'Mother of Deep-Sea Fisheries'. This revolutionary design made large scale trawling in the ocean possible for the first time, resulting in a massive migration of fishermen from the ports in the South of England, to villages further north, such as Scarborough, Hull, Grimsby, Harwich and Yarmouth, that were points of access to the large fishing grounds in the Atlantic Ocean.

The small village of Grimsby grew to become the largest fishing port in the world by the mid 19th century. An Act of Parliament was first obtained in 1796, which authorised the construction of new quays and dredging of the Haven to make it deeper. It was only in the 1846, with the tremendous expansion in the fishing industry, that the Grimsby Dock Company was formed. The foundation stone for the Royal Dock was laid by Albert the Prince consort in 1849. The dock covered 25 acres (10 ha) and was formally opened by Queen Victoria in 1854 as the first modern fishing port. The elegant Brixham trawler spread across the world, influencing fishing fleets everywhere. By the end of the 19th century, there were over 3,000 fishing trawlers in commission in Britain, with almost 1,000 at Grimsby. These trawlers were sold to fishermen around Europe, including from the Netherlands and Scandinavia. Twelve trawlers went on to form the nucleus of the German fishing fleet.

FURTHER DEVELOPMENT

The earliest steam powered fishing boats first appeared in the 1870s and used the trawl system of fishing as well as lines and drift nets. These were large boats, usually 80–90 feet (24–27 m) in length with a beam of around 20 feet (6. 1 m). They weighed 40-50 tons and travelled at 9–11 knots (17–20 km/h; 10–13 mph). The earliest purpose built fishing vessels were designed and made by David Allan in Leith, Scotland in March 1875, when he converted a drifter to steam power. In 1877, he built the first screw propelledsteam trawler in the world.

Steam trawlers were introduced at Grimsby and Hull in the 1880s. In 1890 it was estimated that there were 20,000 men on the North Sea. The steam drifter was not used in the herring fishery until 1897. The last sailing fishing trawler was built in 1925 in Grimsby. Trawler designs adapted as the way they were powered changed from sail to coal-fired steam by World War I to diesel and turbines by the end of World War II.

In 1931, the first powered drum was created by Laurie Jarelainen. The drum was a circular device that was set to the side of the boat and would draw in the nets. Since World War II, radio navigation aids and fish finders have been widely used. The first trawlers fished over the side, rather than over the stern. The first purpose built stern trawler wasFairtry built in 1953 at Aberdeen, Scotland. The ship was much larger than any other trawlers then in operation and inaugurated the era of the 'super trawler'. As the ship pulled its nets over the stern, it could lift out a much greater haul of up to 60 tons. The ship served as a basis for the expansion of 'super trawlers' around the world in the following decades.

RECREATIONAL FISHING

The early evolution of fishing as recreation is not clear. For example, there is anecdotal evidence for fly fishing in Japan, however, fly fishing was likely to have been a means of survival, rather than recreation. The earliest English essay on recreational fishing was published in 1496, byDame Juliana Berners, the prioress of the Benedictine Sopwell Nunnery. The essay was titled Treatyse of Fysshynge wyth an Angle, and included detailed information on fishing waters, the construction of rods and lines, and the use of natural baits and artificial flies.

Recreational fishing took a great leap forward after the English Civil War, where a newly found interest in the activity left its mark on the many books and treatises that were written on the subject at the time. Compleat Angler was written by Izaak Walton in 1653 (although Walton continued to add to it for a quarter of a century) and described the fishing in the Derbyshire Wye. It was a celebration of the art and spirit of fishing in prose and verse. A second part to the book was added by Walton's friend Charles Cotton. Charles Kirby designed an improved fishing hook in 1655 that remains relatively unchanged to this day. He went on to invent the Kirby bend, a distinctive hook with an offset point, still commonly used today.

The 18th century was mainly an era of consolidation of the techniques developed in the previous century. Running rings began to appear along the fishing rods, which gave anglers greater control over the cast line. The rods themselves were also becoming increasingly sophisticated and specialized for different roles. Jointed rods became common from the middle of the century and bamboo came to be used for the top section of the rod, giving it a much greater strength and flexibility.

The industry also became commercialized - rods and tackle were sold at the haberdashers store. After the Great Fire of London in 1666, artisans moved to Redditch which became a centre of production of fishing related products from the 1730s. Onesimus Ustonson established his trading shop in 1761, and his establishment remained as a market leader for the next century. He received a Royal Warrant and became the official supplier of fishing tackle to three successive monarchs starting with King George IV over this period. He also invented the multiplying winch. The commercialization of the industry came at a time of expanded interest in fishing as a recreational hobby for members of the aristocracy.

The impact of the Industrial Revolution was first felt in the manufacture of fly lines. Instead of anglers twisting their own lines - a laborious and time-consuming process - the new textile spinning machines allowed for a variety of tapered lines to be easily manufactured and marketed. British fly-fishing continued to develop in the 19th Century, with the emergence of fly fishing clubs, along with the appearance of several books on the subject of fly tying and fly fishing techniques. By the mid to late 19th century, expanding leisure opportunities for the middle and lower classes began to have its effect on fly fishing, which steadily grew in mass appeal. The expansion of the railway network in Britain allowed the less affluent for the first time to take weekend trips to the seaside or to rivers for fishing. Richer hobbyists ventured further abroad. The large rivers of Norway replete with large stocks of salmon began to attract fishers from England in large numbers in the middle of the century - Jones's guide to Norway, and salmon-fisher's pocket companion, published in 1848, was written by Frederic Tolfrey and was a popular guide to the country.

TECHNOLOGICAL IMPROVEMENTS

Modern reel design had begun in England during the latter part of the 18th century, and the predominant model in use was known as the 'Nottingham reel'. The reel was a wide drum which spooled out freely, and was ideal for allowing the bait to drift along way out with the current. Geared multiplying reels never successfully caught on in Britain, but had more success in the United States, where similar models were modified by George Snyder of Kentucky into his bait-casting reel, the first American-made design in 1810.

The material used for the rod itself changed from the heavy woods native to England, to lighter and more elastic varieties imported from abroad, especially from South America and the West Indies. Bamboo rods became the generally favoured option from the mid 19th century, and several strips of the material were cut from the cane, milled into shape, and then glued together to form light, strong, hexagonal rods with a solid core that were superior to anything that preceded them. George Cotton and his predecessors fished their flies with long rods, and light lines allowing the wind to do most of the work of getting the fly to the fish.

Tackle design began to improve from the 1880s. The introduction of new woods to the manufacture of fly rods made it possible to cast flies into the wind on silk lines, instead of horse hair. These lines allowed for a much greater casting distance. However, these early fly lines proved troublesome as they had to be coated with various dressings to make them float and needed to be taken off the reel and dried every four hours or so to prevent them from becoming waterlogged. Another negative consequence was that it became easy for the much longer line to get into a tangle - this was called a 'tangle' in Britain, and a 'backlash' in the US. This problem spurred the invention of the regulator to evenly spool the line out and prevent tangling.

The American, Charles F. Orvis, designed and distributed a novel reel and fly design in 1874, described by reel historian Jim Brown as the "benchmark of American reel design," and the first fully modern fly reel. Albert Illingworth, 1st Baron Illingworth a textiles magnate, patented the modern form of fixed-spool spinning reel in 1905. When casting Illingworth's reel design, the line was drawn off the leading edge of the spool, but was restrained and rewound by a line pickup, a device which orbits around the stationary spool. Because the line did not have to pull against a rotating spool, much lighter lures could be cast than with conventional reels. The development of inexpensive fiberglass rods, synthetic fly lines, and monofilament leaders in the early 1950s, that revived the popularity of fly fishing.

FISH CAPTURE TECHNOLOGY

Fish capture technology encompasses the process of catching any aquatic animal, using any kind of fishing methods, often operated from a vessel. Use of fishing methods varies, depending on the types of fisheries, and can range from a simple and small hook attached to a line to large and sophisticated midwater trawls or purse seines operated by large fishing vessels. The targets of capture fisheries can include aquatic organisms from small invertebrates to large tunas and whales, which might be found anywhere from the ocean surface to 2 000 meters deep. The large diversity of target species in capture fisheries and their wide distribution requires a variety of fishing gear and methods for efficient harvest. These technologies have developed around the world according to local traditions and, not least, technological advances in various diciplines.

In recent decades major improvements in fibre technology, along with the introduction of other modern materials, have made possible, for example, changes in the design and size of fishing nets. The mechanization of gear handling has vastly expanded the scale on which fishing operations can take place. Improved vessel and gear designs, using computer-aided design methods, have increased the general economics of fishing operations. The development of electronic instruments and fish detection equipment has led to the more rapid location of fish and the lowering of the unit costs of harvesting, particularly as this equipment becomes more widespread. Developments in refrigeration, ice-making and fish processing equipment have contributed to the design of vessels capable of remaining at sea for extended periods.

Although these technologies are largely available, those actually introduced in many small-scale fisheries may amount to no more than motorizing a dugout canoe, use of modern and lighter gear or introducing the use of iceboxes to ensure the quality of the product landed. The impact of such changes, however, has considerably increased landings and the earnings of fishers, and underlines the need for effective management to prevent excessive fishing effort. The emphasis of much recent technical innovation has been focused on greater and more appropriate selectivity of fishing gear so as to reduce negative impacts on the environment.

TYPES OF FISHERIES

Capture fisheries are extremely diversified, comprising a large number of types of fisheries that are categorized by different levels of classification. On a broad level, capture fisheries can be classified as industrial, small-scale/artisanal and recreational. A more specific level includes reference to the fishing area, gear and the main target species, such as the North Sea herring purse seine fishery, Gulf of Mexico shrimp trawl fishery, southern ocean Patagonian toothfish longline fishery. While capture fisheries encompass thousands of fisheries on a global scale, they are often categorized by the capture species, the fishing gear used and the level at which a fishery is managed nationally and/or regionally. The following brief descriptions provide an overview of capture fishery types.

INDUSTRIAL FISHERIES

Capital-intensive fisheries using relatively large vessels with a high degree of mechanization and that normally have advanced fish finding and navigational equipment. Such fisheries have a high production capacity and the catch per unit effort is normally relatively high. In some areas of the world, the term "industrial fisheries" is synonymous with fisheries for species that are used for reduction to fishmeal and fish oil (*e. g.* , the trawl fishery for sandeel in the North Sea or the Peruvian ourse-seine fishery for anchoveta).

SMALL-SCALE FISHERIES

Labour-intensive fisheries using relatively small crafts (if any) and little capital and equipment per person-on-board. Most often family-owned. May be commercial or for subsistence. Usually low fuel consumption. Often equated with artisanal fisheries.

ARTISANAL FISHERIES

Typically traditional fisheries involving fishing households (as opposed to commercial companies), using relatively small amount of capital, relatively small fishing vessels, making short fishing trips, close to shore, mainly for local consumption. In practice, definition varies between countries, *e. g.* , from hand-collection on the beach or a one-person canoe in poor developing countries, to more than 20 m. trawlers, seiners, or long-liners over 20m in developed countries. Artisanal fisheries can be subsistence or commercial fisheries, providing for local consumption or export. Sometimes referred to as small-scale fisheries In general, though by no means always, using relatively low level technology. Artisanal and industrial fisheries frequently target the same resources that may give rise to conflict.

Fig. Artisanal fishing.

RECREATIONAL (SPORT) FISHERIES

Harvesting fish for personal use, leisure, and challenge (*e. g.* , as opposed to profit or research). Recreational fishing does not include sale, barter or trade of all or part of the catch.

COMMERCIAL FISHERIES

Fisheries undertaken for profit and with the objective to sell the harvest on the market, through auction halls, direct contracts, or other forms of trade.

SUBSISTENCE FISHERIES

A fishery where the fish caught are shared and consumed directly by the families and kin of the fishers rather than being bought by intermediaries and sold at the next larger market. Pure subsistence fisheries are rare as part of the products are often sold or exchanged for other goods or services

TRADITIONAL FISHERIES

Fisheries established long ago, usually by specific communities that have developed customary patterns of rules and operations. Traditional fisheries reflect cultural traits and attitudes and may be strongly influenced by religious practices or social customs. Knowledge is transmitted between generations by word of mouth. They are usually small-scale and/or artisanal.

TECHNIQUES

Fig. Fishermen with traditional fish traps, Vietnam.

There are many fishing techniques and tactics for catching fish. The term can also be applied to methods for catching other aquatic animals such as molluscs (shellfish, squid, octopus) and edible marine invertebrates. Fishing techniques include hand gathering, spearfishing, netting, angling and trapping. Recreational, commercial and artisanal fishers use different techniques, and also, sometimes, the same techniques. Recreational fishers fish for pleasure or sport, while commercial fishers fish for profit. Artisanal fishers use traditional, low-tech methods, for survival in third-world countries, and as a cultural heritage in other countries. Usually, recreational fishers use angling methods and commercial fishers use netting methods.

There is an intricate link between various fishing techniques and knowledge about the fish and their behaviour including migration,foraging and habitat. The effective use of fishing techniques often depends on this additional knowledge. Some fishermen followfishing folklores which claim that fish feeding patterns are influenced by the position of the sun and the moon.

FISHING TECHNIQUES

Fishing without gear

Fishing without gear is the simplest form of fishing and is also the oldest. This method is also called collecting by hand, and is still practiced today by professional and non-professional fishermen alike. Most fishing without gear is done during low tide in shallow water and sometimes in deeper water with or without diving suits. Sometimes primitive tools such as hoes or picks are used for collecting the aquatic animals, but they are not classified as gear

GRAPPLING AND WOUNDING GEAR

This technique involves the use of tools such as spears for catching fish and the use of rakes, clamps or tongs to collect shellfish. Other tools such as bows, guns, blowpipes, and other projective devices fall in this category

STUNNING

This technique involves stunning fish with such things as chemicals, explosions or electric shocks. After being stunned the helpless fish are collected. This technique is not considered environmentally sound and has caused extensive damage to coral reefs and is commonly outlawed.

LINE FISHING

Recreational fishing involving a fishing rod is a form of line fishing. In line fishing, a bait or lure is placed on a line. Line fishing can involve one lure placed on a line or thousands, depending upon the harvest size. The most common form of commercial line fishing is long lining where many hooks are used in combination with power winches.

TRAPPING

Trapping involves using some form of device that traps the intended species in it for later retrieval. Mechanical traps that are triggered by the entrance of the intended prey are rarely used by fishermen. Usually the trap uses a labyrinth or other retarding devices to prevent the animal from escaping. The diagram to the right illustrates a form of cage trapping. The most common use of this techniques are lobster and crab pots.

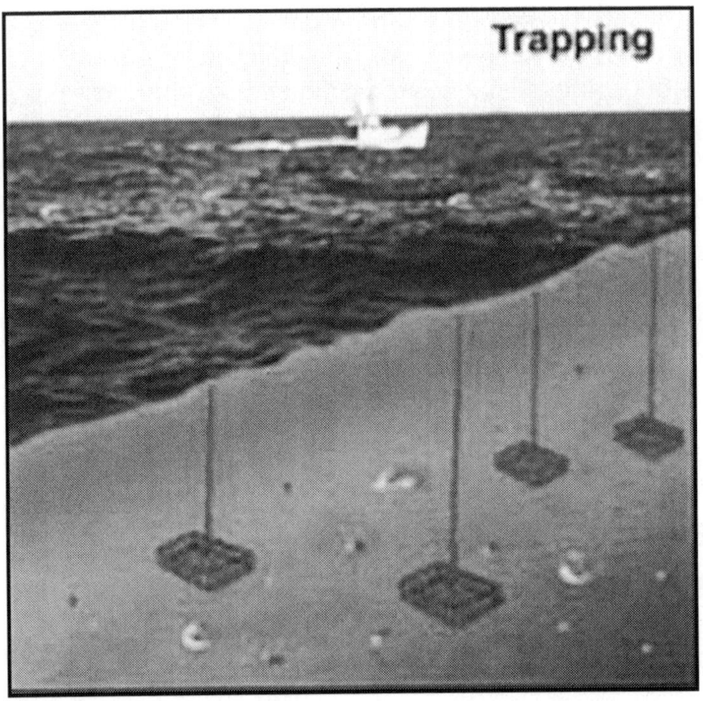

TRAPPING IN THE AIR

Trapping in the air is another form of trapping, except the traps are placed above the water's surface to catch animals such as shrimp or flying fish. The fish are stirred up and when they try to escape they fly into the nets.

FISHING WITH BAG NETS

Bag nets are containers made of nets that are dragged through the water, collecting the fish in its path. The nets are usually held open by a frame and the water current. Below is a diagram of what a bag net looks like and how it is used.

DREDGING AND TRAWLING

This technique is used mainly for shellfish and other organisms such as shrimp, groundfish such as cod and silver hake (whiting), and cephalopods such as squid, that reside on the ocean floor. The design of the dredge or trawl varies depending upon the species being harvested. These tools drag across the sea floor collecting everything in its path. This is one of the most important fishing techniques, second only to seining in total catches

SEINING

Sein nets consist of a single net with two ends that is dragged through the water, collecting fish in its path. Purse seining is very similar to seining. Purse seining involves laying a long open ended net in circular fashion in the water and then carefully closing the bottom end of the net trapping the fish in the net which is then pulled out of the water.

FISHING WITH SURROUNDING NETS

In this system the fish are surrounded by nets and trapped. The trapped fish are then pulled aboard the ship in the nets. Pollock are harvested using this technique.

DRIVING FISH INTO A NET

This practice is similar to the idea of stampeding a herd of cattle or horse off a cliff. In a similar fashion fish are herded together and trapped in a net.

FISHING WITH LIFT NETS

Lift nets consist of nets that are lowered into the water and then as they are lifted up out of the water they catch the fish or crustaceans swimming above it.

FISHING WITH FALLING GEAR

Falling gear uses the same principle as lift nets. Using the falling gear, the nets or baskets are dropped into the water and trap everything underneath them.

GILL NETTING

Gill netting involves using nets that are specially designed such that when the fish intended to be captured swims into it, its gills become caught in the net. These nets are usually allowed to drift freely as the target fish become ensnared in them. By specially designing the nets, most of the capture can be limited to the target species, however larger species can still get caught in the nets.

FISHING WITH ENTANGLED NETS

Entangling nets work be trapping fish in the net, causing them to become tangled in such a way that they cannot escape. Often multiple layers of nets are used to entangle the aquatic animals.

HARVESTING MACHINES

Harvesting machines are a relatively new technique. Harvesting machines involve pumps that pump the fish out of the sea, and other tools such as mechanized dredges that dig up underground mollusks and then transport them to the surface.

TACKLE

Fig. An angler on the Kennet and Avon Canal, England, with his tackle.

Fishing tackle is a general term that refers to the equipment used by fishermen when fishing. Almost any equipment or gear used for fishing can be called fishing tackle. Some examples are hooks, lines, sinkers, floats, rods, reels,baits, lures, spears, nets, gaffs, traps, waders and tackle boxes. Tackle that is attached to the end of a fishing line is called terminal tackle. This includes hooks, sinkers, floats, leaders, swivels, split rings and wire, snaps, beads, spoons, blades, spinners and clevises to attach spinner blades to fishing lures. Fishing tackle can be contrasted with fishing techniques. Fishing tackle refers to the physical equipment that is used when fishing, whereas fishing techniques refers to the ways the tackle is used when fishing.

HOOK, LINE AND SINKER

HOOKS

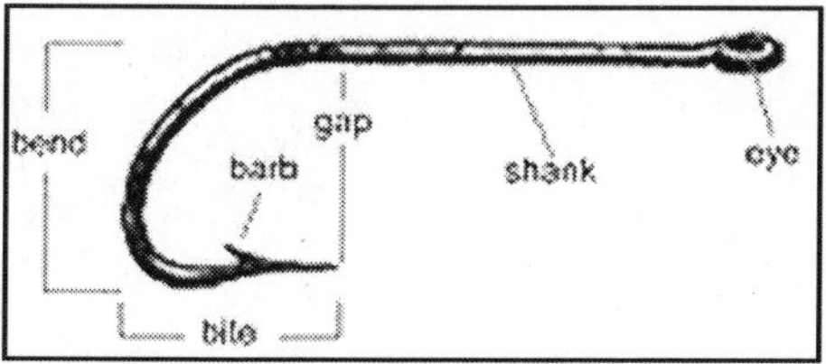

Fig. A fish hook.

The use of the hook in angling is descended, historically, from what would today be called a "gorge". The word "gorge", in this context, comes from an archaic word meaning "throat". Gorges were used by ancient peoples to capture fish. A gorge was a long, thin piece of bone or stone attached by its midpoint to a thin line. The gorge would be fixed with a bait so that it would rest parallel to the lay of the line. When a fish swallowed the bait, a tug on the line caused the gorge to orient itself at right angles to the line, thereby sticking in the fish's gullet. A fish hook is a device for catching fish either by impaling them in the mouth or, more rarely, by snagging the body of the fish. Fish hooks have been employed for millennia by fishermen to catch fresh and saltwater fish. Early hooks were made from the upper bills of eagles and from bones, shells, horns and thorns of plants (Parker 2002). In 2005, the fish hook was chosen by Forbes as one of the top twenty tools in the history of man. Fish hooks are normally attached to some form of line or lure device which connects the caught fish to the fisherman.

There is an enormous variety of fish hooks. Sizes, designs, shapes, and materials are all variable depending on the intended purpose of the hook. They are manufactured for a range of purposes from general fishing to extremely limited and specialized applications. Fish hooks are designed to hold various types of artificial, processed, dead or live baits (bait fishing); to act as the foundation for artificial representations of fish prey (fly fishing); or to be attached to or integrated into other devices that represent fish prey (lure fishing).

LINES

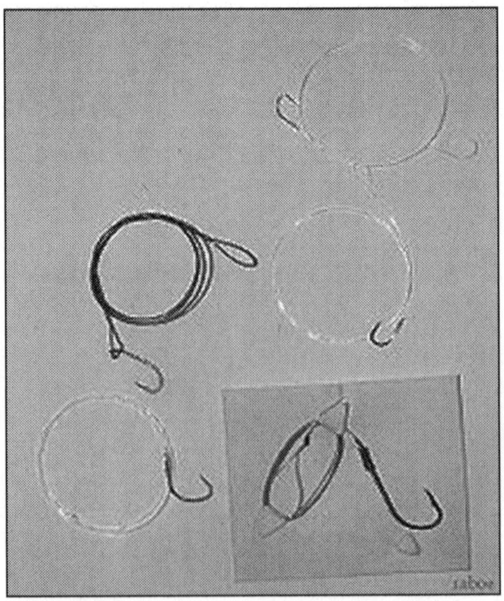

Fig. Fishing line with hooks attached.

A fishing line is a cord used or made for fishing. The earliest fishing lines were made from leaves or plant stalk (Parker 2002). Later lines were constructed from horse hair or silk thread, with catgut leaders. From the 1850s, modern industrial machinery was employed to fashion fishing lines in quantity. Most of these lines were made from linen or silk, and more rarely cotton.

Modern lines are made from artificial substances, including nylon, polyethylene, dacron and dyneema. The most common type ismonofilament made of a single strand. Fishermen often use monofilament because of its buoyant characteristics and its ability to stretch under load. Recently, other alternatives to standard nylon monofilament lines have been introduced made of copolymers or fluorocarbon, or a combination of the two materials. There are also braided fishing lines, cofilament and thermally fused lines, also known as 'superlines' for their small diameter, lack of stretch, and great strength relative to standard nylon monofilament lines.

Important parameters of a fishing line are its length, material, and weight (thicker, sturdier lines are more visible to fish). Factors that may determine what line an angler chooses for a given fishing environment include breaking strength, knot strength, UV resistance, castability, limpness, stretch, abrasion resistance, and visibility.

Fishing with a hook and line is called angling. In addition to the use of the hook and line used to catch a fish, a heavy fish may be landed by using a landing net or a hooked pole called a gaff. Trolling is a technique in which a fishing lure on a line is drawn through the water. Snagging is a technique where the object is to hook the fish in the body.

SINKERS

Fig. Three types of small lead sinkers.

A sinker or plummet is a weight used when angling to force the lure or bait to sink more rapidly or to increase the distance that it may be cast. The ordinary plain sinker is traditionally made of lead. It can be practically any shape, and is often shaped round like a pipe-stem, with a swelling in the middle. However, the use of smaller lead based fishing sinkers has now been banned in the UK, Canada and some states in the USA, since lead can cause toxic lead poisoning if ingested. There are loops of brass wire on either end of the sinker to attach the line. Weights can range from a quarter of an ounce for trout fishing up to a couple of pounds or more for sea bass and menhaden.

The swivel sinker is similar to the plain one, except that instead of loops, there are swivels on each end to attach the line. This is a decided improvement, as it prevents the line from twisting and tangling. In trolling, swivel sinkers are indispensable. The slide sinker, for bottom fishing, is a leaden tube which allows the line to slip through it, when the fish bites. This is an excellent arrangement, as the fisherman can feel the smallest bite, whereas in the other case the fish must first move the sinker before the fisherman feels him.

FISHING RODS

Fig. Fishing with a fishing rod.

A fishing rod is an additional tool used with the hook, line and sinker. A length of fishing line is attached to a long, flexible rod or pole: one end terminates with the hook for catching the fish. Early fishing rods are depicted on inscriptions in ancient Egypt, China, Greece andRome. In Medieval England they were called angles (hence the term angling). As they evolved they were made from materials such as split Tonkin bamboo, Calcutta reed, or ash wood, which were light, tough, and pliable. The butts were frequently made of maple. Handles and grips were made of cork, wood, or wrapped cane. Guides were simple wire loops.

Modern rods are sophisticated casting tools fitted with line guides and a reel for line stowage. They are most commonly made offibreglass, carbon fibre or, classically, bamboo. Fishing rods vary in action as well as length, and can be found in sizes between 24 inches and 20 feet. The longer the rod, the greater the mechanical advantage in casting. There are many different types of rods, such as fly rods, tenkara rods, spin and bait casting rods, spinning rods, ice rods, surf rods, sea rods and trolling rods.

Fishing rods can be contrasted with fishing poles. A fishing pole is a simple pole or stick with a line which is fastened to the tip and suspended with a hooked lure or bait at the other end.

FISHING REELS

Fig. A spinning reel.

A fishing reel is a device used for the deployment and retrieval of a fishing line using a spool mounted on an axle. Fishing reels are traditionally used in angling. They are most often used in conjunction with a fishing rod, though some specialized reels are mounted oncrossbows or to boat gunwales or transoms. The earliest known illustration of a fishing reel is from Chinese paintings and records beginning about 1195 A. D. Fishing reels first appeared in England around 1650 A. D. , and by the 1760s, London tackle shops were advertising multiplying or gear-retrieved reels. Paris, Kentucky native George Snyder is generally given credit for inventing the first fishing reel in America around 1820, a bait casting design that quickly became popular with American anglers.

FISHING BAIT

NATURAL BAITS

Fig. Green Highlander, an artificial fly used for salmon fishing.

The natural bait angler usually uses a common prey species of the fish as an attractant. The natural bait used may be alive or dead. Common natural baits include bait fish, worms, leeches, minnows, frogs, salamanders, nightcrawlers and other insects. Natural baits are effective due to the lifelike texture, odour and colour of the bait presented.

The common earthworm is a universal bait for fresh water angling. In the quest for quality worms, some fishers culture their own worm compost or practice worm charming. Grubs and maggots are also considered excellent bait when trout fishing. Grasshoppers, flies, beesand even ants are also used as bait for trout in their season, although many anglers believe that trout or salmon roe is superior to any other bait. Studies show that natural baits like croaker and shrimp are more recognized by the fish and are more readily accepted. A good bait for red drum is menhaden. Because of the risk of transmitting whirling disease, trout and salmon should not be used as bait. Processed baits, such as groundbait and boilies, can work well with coarse fish, such as carp. For example, in lakes in southern climates such as Florida, fish such as bream will take bread bait. Bread bait is a small amount of bread, often moistened by saliva, balled up to a small size that is bite size to small fish.

ARTIFICIAL BAITS

Many people prefer to fish solely with lures, which are artificial baits designed to entice fish to strike. The artificial bait angler uses a man-made lure that may or may not represent prey. The lure may require a specialised presentation to impart an enticing action as, for example, in fly fishing. Recently, electronic lures have been developed to attract fish. Fishermen have also begun using plastic bait. A common way to fish a soft plastic worm is the Texas rig.

BITE INDICATORS

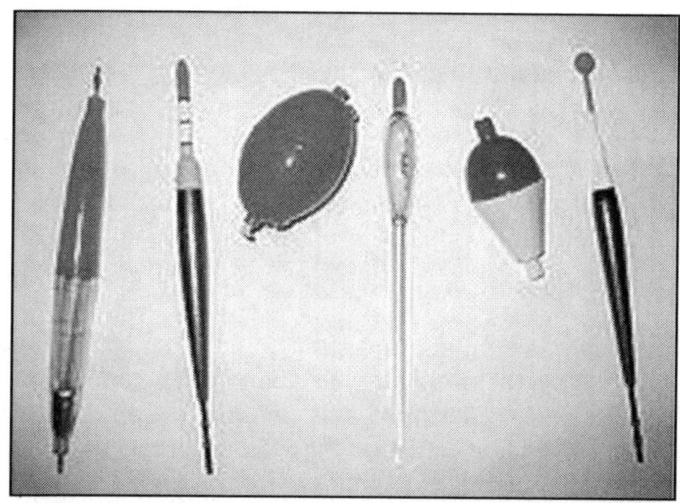

Fig. Different types of fishing floats.

A bite indicator or commonly referred to as "strike indicator" is a mechanical or electronic device which indicates to an angler that something is happening at the hook end of the fishing line. There are many types of bite indicators. Which ones work best depends on the type of fishing. Strike indicators are also commonly used in fly fishing for trout. Nymphs or wet flies will be suspended below the indicator to alert the angler of a fish. This has been used successfully for all trout and salmon species.

Another common form of using an indicator in fly fishing is the "hopper dropper rig" With this technique a nymph or wet fly will be hung from the bottom of a floating dry fly. This will effectively double the opportunity of a strike from an actively feeding fish. Other devices which are widely used as bite indicators are floats which float in the water, and dart about if a fish bites, and quiver tipswhich are mounted onto the tip of the fishing rod. Bite alarms are electronic devices which bleep when a fish tugs a fishing line. Whereas floats and quiver tips are used as visual bite detectors, bite alarms are audible bite detectors.

SPEARS

Spearfishing is an ancient method of fishing conducted with an ordinary spear or a specialised variant such as a harpoon, trident, arrowor eel spear. Harpoons are spears which have a barb at the end. Their use was widespread in palaeolithic times. Cosquer cave in Southern France contains cave art over 16,000 years old, including drawings of seals which appear to have been harpooned. Tridents are spears which have three prongs at the business end. They are also called leisters or gigs. They feature widely in early mythology and history. Modern spears can be used with a speargun. Some spearguns use slings (or rubber loops) to propel the spear. Polespears have a sling attached to the spear, Hawaiian slings have a sling separate from the spear, in the manner of an underwater bow and arrow. A bow or crossbow can be used with arrows in bowfishing.

FISH STRINGER

A fish stringer is a line of rope or chain along which a fisherman can string fish that have been caught so they can be immersed and kept alive in water.

FLY FISHING TACKLE

Fly fishing tackle is equipment used by, and often specialised for use by fly anglers. Fly fishing tackle includes fly lines designed for easy casting, specialised fly reels designed to hold a fly line and supply drag if required for landing heavy or fast fish, specialised fly rods designed to cast fly lines and artificial flies, terminal tackle including artificial flies, and other accessories including fly boxes used to store and carry artificial flies.

TACKLE BOXES

Fig. Typical tackle box with rod and bait bucket.

Fishing tackle boxes have for many years been an essential part of the anglers equipment. Fishing tackle boxes were originally made of wood or wicker and eventually some metal fishing tackle boxes were manufactured. The first plastic fishing tackle boxes were manufactured by Plano in response to the need for a product that didn't rust. Early plastic fishing tackle boxes were similar to tool boxes but soon evolved into the hip roof cantilever tackle boxes with numerous small trays for small tackle. These types of tackle boxes are still available today but they have the disadvantage that small tackle gets mixed up. Fishing tackle boxes have also been manufactured so the drawers themselves become small storage boxes, each with their own lids. This prevents small tackle from mixing, and can turn each drawer into a stand-alone container which can be used to carry small tackle to a rod some distance from the main tackle box.

TRADITIONAL FISHING

Traditional fishing is any kind of small scale, commercial or subsistence fishing practices using traditional techniques such as rod and tackle, arrows and harpoons, throw netsand drag nets, etc.

RECREATIONAL FISHING

Recreational and sport fishing are fishing primarily for pleasure or competition. Recreational fishing has conventions, rules, licensing restrictions and laws that limit the way in which fish may be caught; typically, these prohibit the use of nets and the catching of fish with hooks not in the mouth. The most common form of recreational fishing is done with a rod, reel, line, hooks and any one of a wide range of baits or lures such as artificial flies. The practice of catching or attempting to catch fish with a hook is generally known as angling. In angling, it is sometimes expected or required that fish be returned to the water (catch and release). Recreational or sport fishermen may log their catches or participate in fishing competitions.

Big-game fishing is fishing from boats to catch large open-water species such as tuna, sharks, and marlin. Sport fishing (sometimes game fishing) is recreational fishing where the primary reward is the challenge of finding and catching the fish rather than the culinary or financial value of the fish's flesh. Fish sought after include marlin, tuna, tarpon, sailfish, shark, mackerel, and many others.

FISHING INDUSTRY

The fishing industry includes any industry or activity concerned with taking, culturing, processing, preserving, storing, transporting, marketing or selling fish or fish products. It is defined by the FAO as including recreational, subsistence and commercial fishing, and the harvesting, processing, and marketing sectors. The commercial activity is aimed at the delivery of fish and other seafood products for human consumption or for use as raw material in other industrial processes.

There are three principal industry sectors:

- The commercial sector comprises enterprises and individuals associated with wild-catch or aquaculture resources and the various transformations of those resources into products for sale. It is also referred to as the "seafood industry", although non-food items such as pearls are included among its products.
- The traditional sector comprises enterprises and individuals associated

with fisheries resources from which aboriginal people derive products in accordance with their traditions.

- The recreational sector comprises enterprises and individuals associated for the purpose of recreation, sport or sustenance with fisheries resources from which products are derived that are not for sale.

COMMERCIAL FISHING

Commercial fishing is the capture of fish for commercial purposes. Those who practice it must often pursue fish far into the ocean under adverse conditions. Commercial fishermen harvest almost all aquatic species, from tuna, cod and salmon to shrimp, krill, lobster, clams,squid and crab, in various fisheries for these species. Commercial fishing methods have become very efficient using large nets and sea-going processing factories. Individual fishing quotas and international treaties seek to control the species and quantities caught. A commercial fishing enterprise may vary from one man with a small boat with hand-casting nets or a few pot traps, to a huge fleet oftrawlers processing tons of fish every day. Commercial fishing gear includes weights, nets (*e. g.* , purse seine), seine nets (*e. g.* , beach seine), trawls (*e. g.* , bottom trawl), dredges,hooks and line (*e. g.* , long line and handline), lift nets, gillnets, entangling nets and traps.

According to the Food and Agriculture Organization of the United Nations, total world capture fisheries production in 2000 was 86 million tons (FAO 2002). The top producing countries were, in order, the People's Republic of China (excluding Hong Kong and Taiwan), Peru, Japan, the United States, Chile, Indonesia, Russia, India, Thailand, Norway and Iceland. Those countries accounted for more than half of the world's production; China alone accounted for a third of the world's production. Of that production, over 90% was marine and less than 10% was inland. A small number of species support the majority of the world's fisheries. Some of these species are herring, cod, anchovy, tuna, flounder,mullet, squid, shrimp, salmon, crab, lobster, oyster and scallops. All except these last four provided a worldwide catch of well over amillion tonnes in 1999, with herring and sardines together providing a catch of over 22 million metric tons in 1999. Many other species as well are fished in smaller numbers.

FISH FARMS

Fish farming is the principal form of aquaculture, while other methods may fall under mariculture. It involves raising fish commercially in tanks or enclosures, usually for food. A facility that releases juvenile fish into the wild for recreational fishing or to supplement a species' natural numbers is generally referred to as a fish hatchery. Fish species raised by fish farms include salmon, carp, tilapia, catfish and trout. Increased demands on wild fisheries by commercial fishing has caused widespread overfishing. Fish farming offers an alternative solution to the increasing market demand for fish and fish protein.

FISH PRODUCTS

Fish and fish products are consumed as food all over the world. With other seafoods, it provides the world's prime source of high-quality protein: 14–16 percent of the animal protein consumed worldwide. Over one billion people rely on fish as their primary source of animal protein. Fish and other aquatic organisms are also processed into various food and non-food products, such as sharkskin leather, pigments made from the inky secretions of cuttlefish, isinglass used for the clarification of wine and beer, fish emulsion used as a fertilizer, fish glue, fish oil and fish meal. Fish are also collected live for research or the aquarium trade.

FISHERIES MANAGEMENT

Fig. Fishing down the food web.

Fisheries management draws on fisheries science in order to find ways to protect fishery resources so sustainable exploitation is possible. Modern fisheries management is often referred to as a governmental system of (hopefully appropriate) management rules based on defined objectives and a mix of management means to implement the rules, which are put in place by a system of monitoring control and surveillance.

Fisheries science is the academic discipline of managing and understanding fisheries. It is a multidisciplinary science, which draws on the disciplines of oceanography, marine biology, marine conservation, ecology, population dynamics, economics and management in an attempt to provide an integrated picture of fisheries. In some cases new disciplines have emerged, such as bioeconomics.

SUSTAINABILITY

Issues involved in the long term sustainability of fishing include overfishing, by-catch, marine pollution, environmental effects of fishing, climate change and fish farming. Conservation issues are part of marine conservation, and are addressed in fisheries science programmes. There is a growing gap between how many fish are available to be caught and humanity's desire to catch them, a problem that gets worse as the world population grows.

Similar to other environmental issues, there can be conflict between the fishermen who depend on fishing for their livelihoods and fishery scientists who realise that if future fish populations are to be sustainable then some fisheries must limit fishing or cease operations.

MANAGEMENT PLANS, MEASURES AND STRATEGIES

There is a lot of terminology floating around in fisheries management that, unless clearly understood, can cause further confusion in an already confusing environment. In addition to the words 'goals' and 'operational objectives', the terms management plans, management measures and management strategies will crop up on many occasions in this book and we need to clarify what we mean by each of them and how they differ. The Technical Guidelines on Fisheries Management describe a management plan as "a formal or informal arrangement between a fisheries management authority and interested parties which identifies the partners in the fishery and their respective roles, details the agreed objectives for the fishery and specifies the management rules and regulations which apply to it and provides other details about the fishery which are relevant to the task of the management authority. " A well formulated management plan should be prepared for every fishery and the Code of Conduct states that: "Long-term management objectives should be translated into management actions, formulated as a fishery management plan or other management framework. "

The fisheries policy is translated into goals and the goals into objectives that indicate precisely what is expected to be achieved from the fishery. The objectives are achieved through the implementation of a management strategy which will also be a central element of a management plan. The management strategy is the sum of all the management measures selected to achieve the biological, ecological, economic and social objectives of the fishery. It is possible that in a single species fishery a management strategy could consist of a single management measure, such as a specified total allowable catch (TAC), but in practice the great majority of management strategies consist of a number of management measures, encompassing technical, input and output controls and a system of user rights. An effective management strategy, however, should not contain so many management measures that compliance and enforcement become so difficult as to be practically impossible.

A management measure is the smallest unit of the fishery manager's tool kit and consists of any type of control implemented to contribute to achieving the objectives. Management measures are classified as technical measures, input (effort) and output (catch) controls, and any access rights designed around input and output controls. Technical measures can be sub-divided into regulations on gear-type or gear design and closed areas and closed seasons. A minimum legal mesh size, a seasonal closure of the fishery, a total allowable catch (TAC), a limit on the total number of vessels in a fishery, and a licensing scheme to achieve the limit are all examples of management measures. A substantial part of the book is intended to assist managers in considering and selecting different management measures for a given fishery.

PRIMARY CONSIDERATIONS IN FISHERIES MANAGEMENT

If marine living resources were infinite and indestructible, we could leave people to use and abuse them at will. However, this is not the case and we therefore need to manage fisheries to ensure that the resources are utilised in a sustainable and responsible way, and that the potential benefits are not inefficiently dissipated and possibly totally lost. Fisheries production and yield are constrained by a number of factors which can be classified as biological, ecological and environmental, technological, social and cultural, and economic considerations. There are frequently also considerations imposed by other users of the fishing grounds and neighbouring areas.

Biological Considerations

As living populations or communities, aquatic living resources are capable of on-going renewal through the processes of growth in size and mass of individuals and additions to the population or community through reproduction (leading to what in fisheries is often called 'recruitment'). In a population at equilibrium, the additive processes of growth and reproduction on average equal the loss process of total mortality. In an unexploited population, total mortality consists only of natural mortality, made up of processes such as predation, disease, and death through drastic changes in the environment.

In a fished population, total mortality consists of natural mortality plus fishing mortality, and a primary task of fisheries management is to ensure that fishing mortality does not exceed the amount which the population can withstand, in addition to natural mortality, without undue harm or damage to the sustainability and productivity of the population. This requires not only that the total population is maintained above a certain abundance or biomass, but also that the age structure of the population is maintained in a state in which it is able to maintain the level of reproduction, and hence recruitment, necessary to replenish the losses through mortality. Further, fishing over a long period on selected portions of a stock, for example large individuals or individuals spawning at a specific time or locality within a wider spawning season or range, can reduce the frequency of the particular genetic characteristics giving rise to that feature or behaviour.

This has the effect of reducing the overall genetic diversity of the stock. With reduced genetic diversity, the production potential of the population can be adversely affected and it may also become less resilient to environmental variability and change. Fisheries management needs to be aware of this danger and avoid maintaining such selective pressures over a prolonged period. Achieving an appropriate level and pattern of fishing mortality is hindered substantially by difficulties in estimating population abundance and population dynamics rates and the variability in these rates. The fisheries manager must, however, have sufficient knowledge to make good decisions. The Code of Conduct specifies: "... States... should, inter alia, adopt appropriate measures, based on the best scientific evidence available, which are designed to maintain or restore stocks at levels capable of producing maximum sustainable yield, as qualified by relevant environmental and economic factors... " and further "In implementing the precautionary approach States should take into account, inter alia, uncertainties relating to the size and productivity of the stocks, reference points, stock condition in relation to reference points,... ".

Fishery managers must also respect the stock structure of the resources. Fish populations are frequently made up of a number of different stocks, each of which is genetically largely isolated from the others through behavioural or distributional differences. The different stocks also reflect genetic diversity and if a particular stock is fished to extinction or to very low levels, this genetic diversity may be lost.

The stock will not readily be replenished from other stocks, because of the genetic isolation, and therefore the production it was generating will also be lost, leading to a permanent or at least long-term loss of benefits. Fisheries management should therefore attempt to address each stock separately and to ensure sustainable use of each stock and not just of the population as a whole. In this regard, the Code of Conduct states: "To be effective, fisheries management should be concerned with the whole stock unit over its entire area of distribution and take into account previously agreed management measures established and applied in the same region, all removals and the biological unity and other biological characteristics of the stock. "

Ecological and Environmental Considerations

The abundance and dynamics of a population place an important constraint on fisheries but aquatic populations do not live in isolation. They exist as components of a frequently complex ecosystem, consisting of biological components which may feed on, be fed on by, or compete with a given stock or population. Even those populations which are not directly linked through the food web may indirectly affect each other through their direct interactions with predators, prey or competitors of the other. The physical component of the ecosystem, the water itself, the substrate, inflows of freshwater or nutrients and other non-biological processes may also be very important. Different substrates may be essential for the production of food organisms, for shelter, or as spawning or nursery grounds.

The environment of fish is very rarely static and conditions, particularly of the aquatic environment, can vary substantially over time, from hourly variability, such as the tides, to seasonal variability in, for example, water temperature and currents, to decadal variability as in the occurrence of El Niño events and regime shifts. These changes frequently affect the population dynamics of fish populations, resulting in variability in growth rates, recruitment, natural mortality rates or any combination of these. Such variability can also affect the availability of fish resources to fishing gear, not only affecting the success of the fishing industry, but also the way in which the fishery scientist must interpret catch and catch rate information from the fishery.

Changes in any of the biological, chemical, geological or physical components of the ecosystem can have impacts on the resource population and community. Some of these changes may be beyond human control, such as upwelling processes enriching some coastal ecosystems or large scale temperature anomalies, but they still need to be considered in the management of the resource. Others, such as the destruction of coastal habitats for development, or the direct impact of fishing on the substrate or on other species impacting the resources, are due to human action. In these cases, fisheries management should both take into account their impacts on the resource and, in consultation with other relevant agencies and parties, take steps to minimize their impacts on the fishery ecosystem.

The manager also needs to consider the impact of the fishery on the ecosystem as a whole. There are four types of impact of fisheries on the ecosystem: direct impact on the target species; direct impacts on the bycatch species; indirect impacts on other organisms transmitted through the food chain (*i. e.* , by changing the abundance of predators, prey or competitors of a population); and direct impact of fishing on the physical or chemical environment. The manager needs to be aware of these potential impacts and to use management measures that minimise negative impacts.

The potential to address the ecosystem considerations will vary depending on whether they are caused by or independent of human action, but in both cases the constraints imposed on the resources and the fishery by biological and non-biological ecosystem factors need to be recognised. At the most fundamental level, these factors in combination with the biology of the species determine the maximum abundance, or carrying capacity, and productivity of the resources. Changes in the ecosystem can affect both and, where they are occurring, need to be considered by the fisheries manager. Again, these aspects are dealt with by the Code of Conduct. Amongst several references, the impacts of environmental factors on target stocks and species belonging to the same ecosystem or associated with or dependent upon the target stocks, and assess the relationship among the populations in the ecosystem. "

Technological Considerations

The fishery manager has very little, if any, ability to influence directly the dynamics of the fish populations or communities which support a fishery. In some case, particularly inland waters, there may be opportunities and a desire to undertake stock and habitat enhancement and in some coastal fisheries, habitat destruction may have had an impact on fish production. In the latter case, restoration or stabilization may well be an issue the fisheries manager needs to consider. However, in most fisheries, the only mechanism the fishery manager has to ensure sustainable utilization of the resources is by regulating the quantity of fish caught, when and where they are caught and the size at which they are caught. This can be done through directly regulating the catch taken, by regulating the amount of effort allowed in the fishery, by specifying closed seasons and closed areas and by regulating the type of gear and fishing methods used.

However, there are constraints on how precise the manager can be in setting such regulations. Catch controls are often difficult to monitor and therefore to implement. It is difficult to estimate fishing effort precisely, and normally improving technology and developing skills result in on-going increases in the efficiency of fishing operations, leading to continuing increases in effective effort, unless steps are actively taken to counter these improvements or their consequences. Fishing gear is rarely strongly selective and bycatch of non-target species or unwanted sizes of target species is frequently a problem.

The uncertainties in fisheries management are not just at the level of predicting the status and dynamics of the resources, and uncertainties in the real consequences of implementing fishery measures is also a significant problem to the manager.

A fundamental problem in many fisheries is the existence of too much effort. The presence of excess effort will frequently result in on-going pressure on the fisheries manager to exceed the sustainable fishing mortality on a resource. The social and political pressure to provide employment and opportunities for all those with a stake in the fishery is often hard to resist and readily leads to over-exploitation. The Code of Conduct requires that States take measures to prevent or eliminate excess fishing capacity and such is the global level of concern that the FAO members have agreed on an International Plan of Action (IPOA) for the Management of Fishing Capacity.

Social and Cultural Considerations

Human populations and societies are as dynamic as other biological populations, and social changes take place continuously and on different scales, affected by changes in weather, employment, political circumstances, supply of and demand for fisheries products and other factors. Such changes can affect the appropriateness and effectiveness of management strategies, and therefore need to be considered and accommodated by them. However, again as with biological and technological factors, it can be difficult to identify and quantify the key social and cultural factors influencing fisheries management, generating additional uncertainties for the manager.

A major social constraint in fisheries management is that human societies and behaviour are not easily transformed and fishing families and communities may not be willing to move into other occupations, or away from their normal homes when there is surplus capacity in a fishery, even when their quality of life may be suffering as a result of depleted fish resources. The problem is much worse when there are no other opportunities outside of fisheries in which they could earn a basic living.

Under such circumstances, the political decision to reduce capacity in the fishery is an extremely unattractive option, as the short-term costs of excluding dependent people from the fishery will be much more visible and hence unpopular than a "hands-off" approach which allows the resource and fishery to dwindle in magnitude and quality under sustained excess fishing mortality. Nevertheless, the ecological, economic and social consequences of the latter choice are far more serious in the longer term. This reluctance or inability to take decisions with serious, immediate social consequences for some has been one of the constraints most responsible for over-fishing around the world.

A key requirement for ensuring that social and cultural considerations are properly considered is to involve the interested parties in fisheries management, keeping them well-informed on the management aspects of the fishery and providing them with the opportunity to express their needs and concerns. The Code of Conduct suggests that "the interests of fishers, including those engaged in subsistence, small-scale and artisanal fisheries are taken into account;" and "Within areas under national jurisdiction, States should seek to identify relevant domestic parties having a legitimate interest in the use and management of fisheries resources and establish arrangements for consulting them to gain their collaboration in achieving responsible fisheries".

The relative balance between social and economic considerations in a fishery will depend on the priority given by the appropriate authority to social objectives and economic objectives. Social and economic objectives can conflict: for example it is unlikely that maximising economic efficiency and maximising employment could be simultaneously pursued within a given fishery, and attempting to do so will result in conflict. A common example of such conflicts is that between a commercial fleet pursuing essentially economic objectives and an artisanal fleet fulfilling primarily social objectives, with both having an impact on the same stock, and possibly also interfering with each other's fishing operations. It is important for the management authority to have identified such potential conflicts and to have resolved them, identifying and specifying compromise objectives that achieve general support.

Economic Considerations

In a fishery for which sustainable economic efficiency had been specified as the sole benefit to be extracted and in which optimum circumstances prevailed, market forces could be anticipated to lead to the desired objective of economic efficiency. However, in reality such optimum conditions are rarely if ever found and uncertainties and externalities distort the natural selection of market forces.

Uncertainties include unpredictable variability in resources and other sources of imperfect information, and externalities can include the impacts of other fisheries on the target resources (*e. g.*, taking them as bycatch), subsidies, trade regulations, fiscal regulations and variability in markets and demand. All of these introduce complexity and additional uncertainty into a fishery and, without proper management, will lead to sub-optimum economic performance. It is important for the management authority to consider the broad economic context of a fishery, including relevant macroeconomic factors. As with social considerations, this requires close consultation with the legitimate users who will be the ones most affected by and sensitive to these issues.

At one extreme, although still very common in fisheries especially in many developing countries, are the problems of open access fisheries, in which anyone is allowed entry into a fishery. Under these circumstances, people will continue to enter the fishery until the benefits from fishing are so low as to be unattractive to prospective new entrants. How low this is will depend largely on the availability of other options and in many countries, especially developing countries, such alternatives may be extremely scarce. Even where there are reasonable alternatives, the inevitable result of open access fisheries is dissipation of rent leading to very poor economic efficiency and, unless strong and effective management measures are in place and enforced, to over-exploitation of resources. Such circumstances prevail in many fisheries around the world.

Recognizing this most elementary lesson in fisheries management, the Code of Conduct calls for the adoption of "measures to ensure that no vessel (by which should also be understood no shore-based fisher) be allowed to fish unless so authorised. . ", that "States should ensure that the level of fishing is commensurate with the state of fisheries resources. ", and going further than that, that "Where excess fishing capacity exists, mechanisms should be established to reduce capacity to levels commensurate with the sustainable use of fisheries resources so as to ensure that fisheries operate under economic conditions that promote responsible fisheries. ”

Considerations Imposed by Other Parties

Some offshore fisheries operate in effective isolation from any other users and the regional fisheries organisations charged with their management may be able to manage the fisheries without needing to consider conflicts with or interference from non-fishery users. However, the bulk of global fishery landings come from coastal waters and for many if not most of the fisheries producing these landings, other users are an important consideration and frequently a constraint. Other users of the fishing grounds can include, for example, tourism, conservation, oil and gas extraction, offshore mining and shipping, while use of the intertidal and coastal area can include tourism again, aquaculture and mariculture, coastal zone development for housing, business or industry, and agriculture. All of these can impose significant constraints on fishing activities and may be impacted by fishing activities.

The manager therefore needs to be aware of such activities and of real or potential impacts in both directions. When developing management strategies and formulating management measures, potential conflicts with other users need to be identified and addressed, and the potential impacts of other users on the efficacy of the management strategy and measures need to be considered. The strategy must be adapted so as to account for and be robust to these impacts. An unavoidable implication of overlapping interests is that the fishery manager, through the management authority, must ensure that suitable structures and mechanisms are put into place for effective communication and decision-making with representatives of the other users.

CULTURAL IMPACT

COMMUNITY

For communities like fishing villages, fisheries provide not only a source of food and work but also a community and culturalidentity.

- *Semantic impact:* A “fishing expedition” is a situation where an interviewer implies he knows more than he actually does in order to trick his target into divulging more information than he wishes to reveal. Other examples of fishing terms that carry a negative connotation are: “fishing for compliments”, “to be fooled hook, line and sinker” (to be fooled beyond merely “taking the bait”), and the internet scam of Phishing in which a third party will duplicate a web site where the user would put sensitive information (such as bank codes).

Fig. An Assamese woman catching with traditional fish catching device made of Bamboo which is known as "Jakoi "

RELIGIOUS

Fishing has had an effect on all major religions, including Islam, Christianity, Buddhism, Jainism, Zoroastrianism,Hinduism, and the various new age religions. Jesus was known to participate in fishing excursions. According to the Roman Catholic faith the first Pope was a fisherman, the Apostle Peter, a number of the miracles, and many parables and stories reported in the Bible involve it. The Pope's traditional vestments include a fish-shaped hat.

HISTORY OF FISHING

Fishing is the activity of catching fish. It is an ancient practice dating back at least 40,000 years. Since the 16th century fishing vessels have been able to cross oceans in pursuit of fish and since the 19th century it has been possible to use larger vessels and in some cases process the fish on board. Fish are normally caught in the wild. Techniques for catching fishinclude hand gathering, spearing, netting, angling and trapping.

The term fishing may be applied to catching other aquatic animals such as shellfish, cephalopods, crustaceans, andechinoderms. The term is not usually applied to catching aquatic mammals, such as whales, where the term whaling is more appropriate, or to farmed fish. In addition to providing food, modern fishing is also a recreational sport. According to FAO statistics, the total number of fishermen and fish farmers is estimated to be 38 million. Fisheries andaquaculture provide direct and indirect employment to over 500 million people. In 2005, the worldwide per capita consumption of fish captured from wild fisheries was 14. 4 kilograms, with an additional 7. 4 kilograms harvested from fish farms.

PREHISTORY

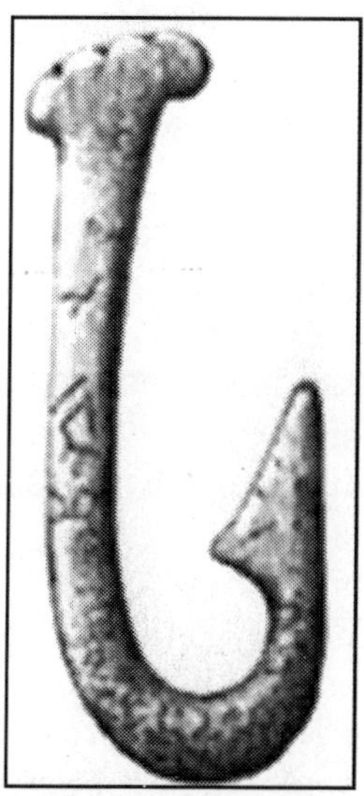

Fig. Stone Agefish hook made from bone.

Fishing is an ancient practice that dates back at least to the Upper Paleolithic period which began about 40,000 years ago. Isotopic analysis of the skeletal remains of Tianyuan man, a 40,000-year-old modern human from eastern Asia, has shown that he regularly consumed freshwater fish. Archaeological features such as shell middens, discarded fish bones and cave paintings show that sea foods were important for survival and consumed in significant quantities. During this period, most people lived a hunter-gatherer lifestyle and were, of necessity, constantly on the move. However, where there are early examples of permanent settlements (though not necessarily permanently occupied) such as those at Lepenski Vir, they are almost always associated with fishing as a major source of food.

Spearfishing with barbed poles (harpoons) was widespread in palaeolithic times. Cosquer cave in Southern France contains cave art over 16,000 years old, including drawings of seals which appear to have been harpooned.

The Neo-lithic culture and technology spread worldwide between 4,000 and 8,000 years ago. With the new technologies of farming and pottery came basic forms of the main fishing methods that are still used today. From 7500 to 3000 years ago, Native Americans of the California coast were known to engage in fishing with gorge hook and line tackle. In addition, some tribes are known to have used plant toxins to induce torpor in stream fish to enable their capture. Copper harpoons were known to the seafaring Harappans well into antiquity. Early hunters in India include the Mincopie people, aboriginal inhabitants of India's Andaman and Nicobar islands, who have used harpoons with long cords for fishing since early times.

EARLY HISTORY

The ancient river Nile was full of fish; fresh and dried fish were a staple food for much of the population. The Egyptians invented various implements and methods for fishing and these are clearly illustrated in tomb scenes, drawings, and papyrus documents. Simple reed boatsserved for fishing. Woven nets, weir baskets made from willow branches, harpoons and hook and line (the hooks having a length of between eight millimetres and eighteen centimetres) were all being used. By the 12th dynasty, metal hooks with barbs were being used. As is fairly common today, the fish were clubbed to death after capture. Nile perch, catfish and eels were among the most important fish. Some representations hint at fishing being pursued as a pastime.

There are numerous references to fishing in ancient literature; in most cases, however, the descriptions of nets and fishing-gear do not go into detail, and the equipment is described in general terms. An early example from theBible in Job 41:7: Canst thou fill his skin with barbed irons? or his head with fish spears? Unlike in Minoan culture, fishing scenes are rarely represented in ancient Greek culture, a reflection of the low social status of fishing. There is a wine cup, dating from c. 500 BC, that shows a boy crouched on a rock with a fishing-rod in his right hand and a basket in his left. In the water below there is a rounded object of the same material with an opening on the top. This has been identified as a fish-cage used for keeping live fish, or as a fish-trap. It is clearly not a net. This object is currently in the Muscum of Fine Arts, Boston.

Oppian of Corycus, a Greek author wrote a major treatise on sea fishing, the Halieulica or Halieutika, composed between 177 and 180. This is the earliest such work to have survived intact to the modern day. Oppian describes various means of fishing including the use of nets cast from boats, scoop nets held open by a hoop, spears and tridents, and various traps "which work while their masters sleep".

Oppian's description of fishing with a "motionless" net is also very interesting: The fishers set up very light nets of buoyant flax and wheel in a circle round about while they violently strike the surface of the sea with their oars and make a din with sweeping blow of poles. At the flashing of the swift oars and the noise the fish bound in terror and rush into the bosom of the net which stands at rest, thinking it to be a shelter: foolish fishes which, frightened by a noise, enter the gates of doom. Then the fishers on either side hasten with the ropes to draw the net ashore.

The Greek historian Polybius (ca 203 BC–120 BC), in his Histories, describes hunting for swordfish by using a harpoon with a barbed and detachable head. Pictorial evidence of Roman fishing comes from mosaics which show fishing from boats with rod and line as well as nets. Various species such as conger, lobster, sea urchin, octopus and cuttlefish are illustrated. In a parody of fishing, a type of gladiator called retiarius was armed with a trident and a casting-net. He would fight against the murmillo, who carried a short sword and a helmet with the image of a fish on the front.

The Greco-Roman sea god Neptune is depicted as wielding a fishing trident. In India, the Pandyas, a classical Dravidian Tamil kingdom, were known for the pearl fishery as early as the 1st century BC. Their seaport Tuticorin was known for deep sea pearl fishing. Theparavas, a Tamil caste centred in Tuticorin, developed a rich community because of their pearl trade, navigation knowledge and fisheries. In Norse mythology the sea giantess Rán uses a fishing net to trap lost sailors. The Moche people of ancient Peru depicted fisherman in their ceramics. From ancient representations and literature it is clear that fishing boats were typically small, lacking a mast or sail, and were only used close to the shore. In traditional Chinese history, history begins with three semi-mystical and legendary individuals who taught the Chinese the arts of civilization around 2800–2600 BC: of these Fu Hsi was reputed to be the inventor of writing, hunting, trapping, and fishing.

Gillnet

Gillnets existed in ancient times as archaeological evidence from the Middle East demonstrates. In North America, aboriginal fishermen used cedar canoes and natural fibre nets, *e. g.* ,, made with nettels or the inner bark of cedar. They would attach stones to the bottom of the nets as weights, and pieces of wood to the top, to use as floats. This allowed the net to suspend straight up and down in the water.

Each net would be suspended either from shore or between two boats. Native fishers in the Pacific Northwest, Canada, and Alaska still commonly use gillnets in their fisheries for salmon and steelhead. Both drift gillnets and setnets also have been widely adapted in cultures around the world. The antiquity of gillnet technology is documented by a number of sources from many countries and cultures. Japanese records trace fisheries exploitation, including gillnetting, for over 3,000 years. Many relevant details are available concerning the Edo period (1603–1867). Fisheries in the Shetland Islands, which were settled by Norsemen during the Viking era, share cultural and technological similarities with Norwegian fisheries, including gillnet fisheries for herring. Many of the Norwegian immigrant fishermen who came to fish in the great Columbia River salmon fishery during the second half of the 19th century did so because they had experience in the gillnet fishery for cod in the waters surrounding the Lofoten Islands of northern Norway. Gillnets were used as part of the seasonal round by Swedish fishermen as well. Welsh and English fishermen gillnetted for Atlantic salmon in the rivers of Wales and England in coracles, using hand-made nets, for at least several centuries. These are but a few of the examples of historic gillnet fisheries around the world.

Cod trade

Fig. Stockfish.

One of the world's longest lasting trade histories is the trade of dry cod from the Lofoten area to the southern parts of Europe, Italy, Spainand Portugal. The trade in cod started during the Viking period or before, has been going on for more than 1000 years and is still important.

Cod has been an important economic commodity in an international market since the Viking period (around 800 AD). Norwegians useddried cod during their travels and soon a dried cod market developed in southern Europe. This market has lasted for more than 1000 years, passing through periods of Black Death, wars and other crises and still is an important Norwegian fish trade. The Portuguesehave been fishing cod in the North Atlantic since the 15th century, and clipfish is widely eaten and appreciated in Portugal. The Basquesalso played an important role in the cod trade and are believed to have found the Canadian fishing banks in the 16th century. The North American east coast developed in part due to the vast amount of cod, and many cities in the New England area spawned near cod fishing grounds.

Apart from the long history this particular trade also differs from most other trade of fish by the location of the fishing grounds, far from large populations and without any domestic market. The large cod fisheries along the coast of North Norway (and in particular close to the Lofoten islands) have been developed almost uniquely for export, depending on sea transport of stockfish over large distances. Since the introduction of salt, dried salt cod ('klippfisk' in Norwegian) has also been exported. The trade operations and the sea transport were by the end of the 14th century taken over by the Hanseatic League, Bergen being the most important port of trade.

William Pitt the Elder, criticizing the Treaty of Paris in Parliament, claimed that cod was "British gold"; and that it was folly to restore Newfoundland fishing rights to the French. In the 17th and 18th centuries, the New World, especially in Massachusetts and Newfoundland, cod became a major commodity, forming trade networks and cross-cultural exchanges.

MODERN TRAWLING

EARLY MODERN DESIGNS

In the 15th century, the Dutch developed a type of seagoing herring drifter that became a blueprint for European fishing boats. This was the Herring Buss, used by Dutch herring fishermen until the early 19th centuries. The ship type buss has a long history. It was known around 1000 AD in Scandinavia as a b□za, a robust variant of the Viking longship. The first herring buss was probably built in Hoornaround 1415. The last one was built in Vlaardingen in 1841.

The ship was about 20 metres long and displaced between 60 and 100 tons. It was a massive round-bilged keel ship with a bluff bow andstern, the latter relatively high, and with a gallery. The busses used long drifting gill nets to catch the herring. The nets would be retrieved at night and the crews of eighteen to thirty men would set to gibbing, salting and barrelling the catch on the broad deck. The ships sailed in fleets of 400 to 500 ships to the Dogger Bank fishing grounds and the Shetland isles. They were usually escorted by naval vessels, because the English considered they were "poaching". The fleet would stay at sea for weeks at a time. The catch would sometimes be transferred to special ships (called ventjagers), and taken home while the fleet would still be at sea (the picture shows aventjager in the distance).

During the 17th century, the British developed the dogger, an early type of sailing trawler or longliner, which commonly operated in theNorth Sea. The dogger takes its name from the Dutch word dogger, meaning a fishing vessel which tows a trawl. Dutch trawling boats were common in the North Sea, and the word dogger was given to the area where they often fished, which became known as the Dogger Bank.

Doggers were slow but sturdy, capable of fishing in the rough conditions of the North Sea. Like the herring buss, they were wide-beamed and bluff-bowed, but considerably smaller, about 15 metres long, a maximum beam of 4. 5 metres, a draught of 1. 5 metres, and displacing about 13 tonnes. They could carry a tonne of bait, three tonnes of salt, half a tonne each of food and firewood for the crew, and return with six tonnes of fish. Decked areas forward and aft probably provided accommodation, storage and a cooking area. An anchor would have allowed extended periods fishing in the same spot, in waters up to 18 metres deep. The dogger would also have carried a small open boat for maintaining lines and rowing ashore.

A precursor to the dory type was the early French bateau type, a flat bottom boat with straight sides used as early as 1671 on the Saint Lawrence River. The common coastal boat of the time was the wherry and the merging of the wherry design with the simplified flat bottom of the bateau resulted in the birth of the dory. Anecdotal evidence exists of much older precursors throughout Europe. England, France, Italy, and Belgium have small boats from medieval periods that could reasonably be construed as predecessors of the Dory.

Dories appeared in New England fishing towns sometime after the early 18th century. They were small, shallow-draft boats, usually about five to seven metres (15 to 22 feet) long. Lightweight and versatile, with high sides, a flat bottom and sharp bows, they were easy and cheap to build. The Banks dories appeared in the 1830s. They were designed to be carried on mother ships and used for fishing cod at the Grand Banks. Adapted almost directly from the low freeboard, French river bateaus, with their straight sides and removable thwarts, bank dories could be nested inside each other and stored on the decks of fishing schooners, such as the Gazela Primeiro, for their trip to the Grand Banks fishing grounds.

2

Fishing Techniques

Fishing techniques are methods for catching fish. The term may also be applied to methods for catching other aquatic animals such as molluscs (shellfish, squid, octopus) and edible marine invertebrates. Fishing techniques include hand gathering, spearfishing, netting, angling and trapping. Recreational, commercial andartisanal fishers use different techniques, and also, sometimes, the same techniques. Recreational fishers fish for pleasure or sport, while commercial fishers fish for profit. Artisanal fishers use traditional, low-tech methods, for survival in third-world countries, and as a cultural heritage in other countries. Mostly, recreational fishers use angling methods and commercial fishers use netting methods.

There is an intricate link between various fishing techniques and knowledge about the fish and their behaviour includingmigration, foraging and habitat. The effective use of fishing techniques often depends on this additional knowledge. Which techniques are appropriate is dictated mainly by the target species and by its habitat. Fishing techniques can be contrasted with fishing tackle. Fishing tackle refers to the physical equipment that is used when fishing, whereas fishing techniques refers to the manner in which the tackle is used when fishing.

FRESHWATER FISHING TECHNIQUES

BAIT CASTING

Bait casting is a style of fishing that relies on the weight of the lure to extend the line into the target area. Bait casting involves a revolving-spool fishing reel (or "free spool") mounted on the topside of the rod. Bait casting is definitely an acquired skill. Once you get the hang of the technique (check out the casting animation), you will be casting your lures right on target into the structures where fish are feeding and hanging out.

With bait casting, you can use larger lures (1/2 to 3/4) and cast them for longer distances. To get started, you'll need a rod with good spring action, a good quality anti-backlash reel, 10 to 15 pound test line and a variety of specific bait casting lures.

SPIN CASTING

We won't say it's foolproof, but spin casting is an ideal fishing method for beginning anglers. Spin-casting equipment is easier to use than bait casting. You can use it to cast both light and heavy lures without tangling or breaking your fishing line. Basic equipment includes a 7-foot rod, a spinning reel and 6 to 10 pound test line for casting 1/16 to 3/4 ounce lures. You can use an open-face, closed-face or spin-cast reel for spin casting. Fishing Tips: How to Cast a Spinning ReelCourtesy of ExpertVillage. com

TROLLING

Trolling is done using a small electric motor that moves the boat quietly through the water so fish aren't spooked. But you can also troll by towing a lure while walking along the edge of a shoreline, bridge or pier. The speed of the boat determines the depth of your bait. And the depth of the bait is determined by the species of fish you're trying to catch. Use a spinning reel or a bait caster for trolling. Some states don't allow motorized trolling, so check out your local fishing regulations to avoid tangling with the fish enforcers.

STILL FISHING

Still fishing is a versatile way to go. You can still fish from a pier, a bridge, an anchored boat or from shore. You can still fish on the bottom or off the bottom in ponds, lakes, rivers and streams for a variety of species. And you can still fish during most seasons and during any part of the day. Your equipment and the size of the hooks and bait you use depends on what kind of fish you're after. But your best equipment for still fishing is patience. You have to wait for the fish to bite.

DRIFT FISHING

Drift fishing allows you to fish over a variety of habitats as your boat drifts with the currents or wind movement. You can drift fish on the bottom or change the depth with a bobber or float. Natural baits work best. But jigs, lures and artificial flies will produce good results, too. You can drift fish on ponds, lakes, rivers and streams any time of the day and year.

LIVE LINING

Your fishing line is "live" when your boat is anchored in a flowing body of water like a river or stream. Use live or prepared fishing bait and keep it on or just off the bottom. Live lining off the bottom allows your line to drift with the current through holes and rocks where the fish may be holding. Your equipment and the size of your fishing hooks and lures depend on what type of fish you're after.

CHUMMING

To attract fish or get them biting again, you can throw "chum" into the water where you're fishing. You can use ground-up bait fish, canned sweet corn, dead minnows in a coffee can for ice fishing, pet food, even breakfast cereal. Or stir up some natural chum by scraping the bottom with a boat oar. Be sure not to over-chum. You want to get them interested in feeding; you do not want to stuff them before they get a chance to go after your hook. Chumming is not legal in all states. Check local fishing fishing regulations to make sure you are not illegally stimulating the hunger of your future catch.

BOTTOM BOUNCING

Bottom Bouncing is done from a drifting or trolling boat, and it's a great way to attract or locate fish during most seasons and times of day. Use a buck tail jig or natural bait and drag it along the bottom. The dragging motion causes the lure to bounce along stirring up small clouds of sand or mud. After a few strikes with bottom bouncing, you can drop anchor and apply other methods to hook the particular kind of species you've attracted.

SURFACE POPPERS

There's nothing quite like the sudden, exciting rush of a fish rising to the surface and exploding onto your lure. Surface poppers are a style of topwater fishing bait that get their action from a cupped face carved or molded into the front of the lure body. Cast your popper out to the target area and let it settle briefly. By taking in small amounts of line slowly, the cupped face "pops" along the surface, imitating the action of prey, such as small insects, small frogs or even a small injured fish. To increase your chances of landing your catch, resist the urge to set the hook immediately when the fish strikes – let it take the popper under the water first – then set your hook firmly.

JIGGING

Jig fishing is popular and challenging. Why? Because the person fishing is creating the action that attracts, or doesn't attract, the particular type of fish he or she is trying to catch. Here's how it works. Cast out and let your jig hook sink to the bottom. Then use your rod tip to raise the bait about a foot off the bottom. Then let it drop back to the bottom. You can jig up and down, side to side or up and down and sideways. Jig rigs come in all sizes, shapes and colours, and can be used with or without live fishing bait.

JIG AND WORM

Attach a worm to your jig hook and use it to bottom hop or sweep through your target area. To bottom hop, cast to the target and let the jig sink. Then reel in slowly, twitching the rod with every third or fourth turn of your reel. To sweep, cast to the target and drag the jig parallel to the bottom while reeling with a fairly tight line. Slow and steady gets the fish when you're sweeping with a jig and worm.

USING SPOONS

Spoons are among the most popular lures and are easy to use. Some are thin and light, some are thick and heavy. And different spoons have different actions. How and where you're fishing will determine how to use them.

- *Casting spoons:* The basic technique is to cast it out and reel it back. A steady retrieve is usually best. If fish are curious but not striking, try slight variations in the speed or direction of your spoon.
- *Trolling spoons:* Thinner and lighter than casting spoons so they can be trolled slowly. Typically used with depth control rig for open water species like trout, salmon or walleye. Can also be tied onto a rig with a diving crankbait and trolled on a long line to go after species near the bottom.
- *Topwater/Weedless spoons:* Great for predators like bass, musky and pike that tend to hide in thick underwater cover. Cast over the cover, start retrieving and reel just fast enough to keep the lure on the surface.
- *Jigging spoons:* Great for predators typically found on deep structure. Let the spoon freefall down. When it hits bottom, take up slack line until the rod tip is a foot above the water, then work the spoon with short jerks up and down. Usually, strikes occur when the spoon is falling, so be ready.

HAND FISHING

It is possible to fish and gather many sea foods with minimal equipment by using the hands. Gathering seafood by hand can be as easy as picking shellfish or kelp up off thebeach, or doing some digging for clams or crabs. The earliest evidence for shellfish gathering dates back to a 300,000-year-old site in France called Terra Amata. This is ahominid site as modern Homo sapiens did not appear until around 50,000 years ago.

- Flounder tramping - Every August, the small Scottish village of Palnackie hosts the world flounder tramping championships where flounder are captured by stepping on them.
- *Noodling:* Is practiced in the United States. The noodler places his hand inside a catfish hole. If all goes as planned, the catfish swims forward and latches onto the noodler's hand, and can then be dragged out of the hole, albeit with risk of injury to the noodler.
- Pearl divers - traditionally harvested oysters by free-diving to depths of thirty metres. Today, free-diving recreational fishers catchlobster and abalone by hand.

Fig. Ama diver in Japan.

- Trout binning - is another method of taking trout. Rocks in a rocky stream are struck with a sledgehammer. The force of the blow stuns the fish.
- Trout tickling - In the British Isles, the practice of catching trout by hand is known as trout tickling; it is an art mentioned several times in the plays of Shakespeare.

SALT WATER HANDLINING

Ocean handlining is often used to catch groundfish and squid but other species are sometimes caught, including pelagic fish. Sea handlining a good way to catch larger oceanic fish.

FRESHWATER HANDLINING

Handlining is also used for catching fresh water fish. Panfish, walleyes, and other freshwater game fish can be caught using handlining fishing techniques. Handlining can be practiced from the shore or from a fishing boat. Walleye anglers practice handlining over moderately deep water in a drifting boat. Handlining is also practiced by ice fishinganglers.

HANDLINING TECHNIQUES

A jigging motion can be used to attract fish which are normally caught while trying to strike the lure but they can also be snagged by the hooks as they investigate the jigged lure. The lure can also be fished motionless and the angler feels for the bait to be picked up by a fish and then sets the hook after waiting for the fish to fully take the bait. After a strike occurs the hook is set and then the fish is hauled in and the caught fish is removed.

SOME SIMPLE TIPS FOR SPEARFISHING

Not everybody is a great spearfisherman, but with a little knowledge and guide from others you can start to increase your skills and catch larger fish. To be a good spearfisherman you need to be a good diver. Improving your diving skills should be your first focus. A good and safe diver can spend more time hunting. Diving skills come with experience. Buoyancy, steady breathing, and other fundamental skills develop with more dives. When you have all of this down you are ready to take on other tasks underwater, like spearfishing.

Second you have to know what you are looking for. After a while, you will get an eye for spotting certain fish you could not before. You will learn to spot fish and recognize common patterns of camouflaged fish. For example the Cabazon. This fish could be two feet from someone and they would never know it. By the time you spot the cabazon he will dart off into the abyss. You will also be able to capitilize off of certain fish instinct and recognize their habitats. There are several common fish such a black or blue rockfish which make great fish tacos but also as an attractive lure to attract larger like fish such as lingcod, sheeps head, vermilion rock fish, Greenling andCabazon. Always keep an eye on these larger fish in the distance and behind you. The lingcod will sometimes stalk you to investigate your catch. Sheeps Head often join to see what commotion is about. I have had both of these fish come right up to the stringer.

The ling cod love to hide out in dens and dark holes. Having a small flashlight, can assist in exploring holes that may hold larger fish. I check every hole I find regardless of size. The biggest fish will often be found here. Cabazon can be found sitting on top of rocks and rocky bottoms. They are masters of camouflage but once you get a eye for them they are easy to spot. Most of the larger fish rest in small or large holes and are nocturnal hunter. A small reef gun capable of reaching this fish can often yield a large fish over 10 lbs.

The sheeps head is a very smart fish. Avoiding eye contact will allow the fish to come closer rather than flee. Gain enough distance and you can position for a more better shot. Make sure you can pop him just behind the head as the head is heavily protected by bone. If you hit him in the head, most of the time the spear will bounce off and scare him away. If it is a large sheeps head make sure the shot is going to penetrate completely. This means big sheeps head need to be at close range with a small gun. I have taken sheeps head in the 35 pound range on a 32 inch gun. Remember this is a smart fish. The sight of a gun to a sheeps head that has been hunted in the past and lived will often avoid you.

The vermilion cod I often find hanging out in palm kelp or between some large bolders to avoid any surge underwater. With this in mind, I will sometimes look on the backside of larger rocks where the water is calm. Most of the time the vermillion hover two or three feet above the bottom and eighty percent of the time they are facing you giving you a small profile to view. Again I try to avoid eye contact. I always try to get a side profile shot in the side right below the eye area. The largest I have seen are around 12-13 pounds.

The greenling is probably one of the most curios fish in the ocean. They are very good eating. Once you have fish on the stringer you will always find one following you around. Identification of this fish is easy. They have a long slender body and have an often yellow green tint. They rarely hide and move a lot. Do not take too long to take your shot as they are fast.

Most of the other Rock fish found in Morro bay can be found hovering over rock and in the same holes as the fish I mentioned above. Remember this is my strategy when I hunt and you might be told by someone else something totally different. But this works for me, I have shown others the same techniques and it seems to work for them. Remember to always check out fishing Regulations on the California Fish & Game web site before going on your next adventure. Be mindful of the fish population and know your fish species. Come join us for a day of spearfishing and share your experience as a hunter under water.

NETTING

Fishing nets are meshes usually formed by knotting a relatively thin thread. About 180 AD the Greek author Oppian wrote the Halieutica, a didactic poem about fishing. He described various means of fishing including the use of nets cast from boats, scoop nets held open by a hoop, and various traps "which work while their masters sleep". Netting is the principal method of commercial fishing, though longlining, trolling, dredging and traps are also used.

Fig. A fisherman casting a net in Kerala, India.

Fig. Oil painting of gillnetting, The salmon fisher by Eilif Peterssen.

Fig. Fishing with nets in Cà Mau Province, Vietnam.

Artisanal Techniques

- Chinese fishing nets - are shore operated lift nets. Huge mechanical contrivances hold out horizontal nets with diameters of twenty metres or more. The nets are dipped into the water and raised again, but otherwise cannot be moved.
- Lampuki nets - are an example of a traditional artisanal use of nets. Since Roman times, Maltese fishers have cut the larger, lowerfronds from palm trees which they then weave into large flat rafts. The rafts are pulled out to sea by a luzzu, a small traditionalfishing boat. In the middle of the day, lampuki fish (the Maltese name for mahi-mahi) school underneath the rafts, seeking the shade, and are caught by the fishers using large mesh nets.
- Cast nets - are round nets with small weights distributed around the edge. They are also called throw nets. The net is caste or thrown by

hand in such a manner that it spreads out on the water and sinks. Fish are caught as the net is hauled back in. This simple device has been in use, with various modifications, for thousands of years.

- Drift nets - are nets which are not anchored. They are usually gillnets, and are commonly used in the coastal waters of many countries. Their use on the high seas is prohibited, but still occurs.
- Ghost nets - are nets that have been lost at sea. They can be a menace to marine life for many years.
- Gillnets - catch fish which try to pass through by snagging on the gill covers. Trapped, the fish can neither advance through the net nor retreat.
- Hand nets - are small nets held open by a hoop. They have been used since antiquity. They are also called scoop nets, and are used for scooping up fish near the surface of the water. They may or may not have a handle–if they have a long handle they are called dip nets. When used by anglers to help land fish they are called landing nets. Because hand netting is not destructive to fish, hand nets are used for tag and release, or capturing aquarium fish.
- Seine nets - are large fishing nets that can be arranged in different ways. In purse seining fishing the net hangs vertically in the water by attaching weights along the bottom edge and floats along the top. Danish seining is a method which has some similarities withtrawling. A simple and commonly used fishing technique is beach seining, where the seine net is operated from the shore.
- Surrounding nets -
- Tangle nets - also known as tooth nets, are similar to gillnets except they have a smaller mesh size designed to catch fish by the teeth or upper jaw bone instead of by the gills.
- Trawl nets - are large nets, conical in shape, designed to be towed in the sea or along the sea bottom. The trawl is pulled through the water by one or more boats, called trawlers. The activity of pulling the trawl through the water is called trawling.

FISHING METHODS

The methods and fishing gear used to catch fish are a determining factor of sustainability as they can impact upon marine life and habitats. Over the past 50 years, fishing technology has advanced greatly, increasing the capacity for boats to locate and catch fish. In this time there have also been advances to reduce the environmental impact of fishing, yet there is room for further improvement and research and development is ongoing. Fishing methods vary in different locations and cultures and the precise impacts of each method are dependent on the robustness of the management and techniques used on individual boats. The main commercial capture methods are described here to help you broadly understand the ways in which fish are caught and the environmental impacts they have.

PURSE SEINE NET

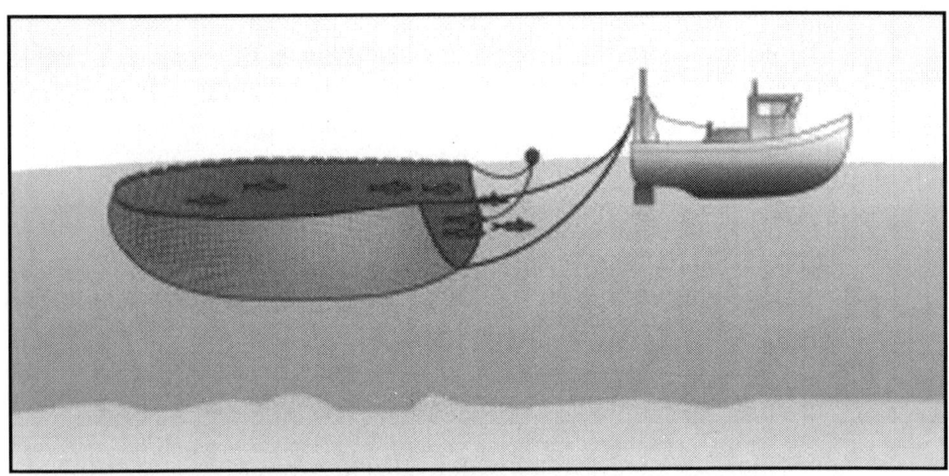

This method of fishing is used to capture large schools of fish at or near the surface. The net itself is in the shape of a long strip or wall and may be up to one kilometer in length and as deep as 200 meters. A system of weights running along the bottom and floats positioned along the top keep the net vertically suspended in the water column. Once a school of fish has been located, which is sometimes done with the help of a spotter aircraft or radar used to detect large flocks of birds feeding at the surface, the net is then pitched into the water and pulled around the school of fish to encircle it in this "wall" of netting.

After the school of fish has been completely surrounded, a rope running along the top of the net, known as the purse line, is pulled to cinch the bottom closed, like a laundry bag.

This conforms the net into a bowl-like shape and the school of fish is pulled on board the fishing vessel. As this technique is employed close to the surface it does come into contact with the ocean floor and therefore does not cause any damage to bottom dwelling (benthic) organisms and habitats. Bycatch is also limited due to the nets being targeted at schools of specific species of fish, however, undesired animals feeding on those schools such as sharks, dolphins, and turtles are sometimes captured. Purse seining is practiced in many commercial fisheries around the globe, and is one of the methods used to harvest Atlantic mackerel, anchovy, sardine, and Alaska salmon which are included on the KidSafe "Best Choices" seafood list. It is also used to catch fish such as bluefin and other tuna species.

TRAWL NET

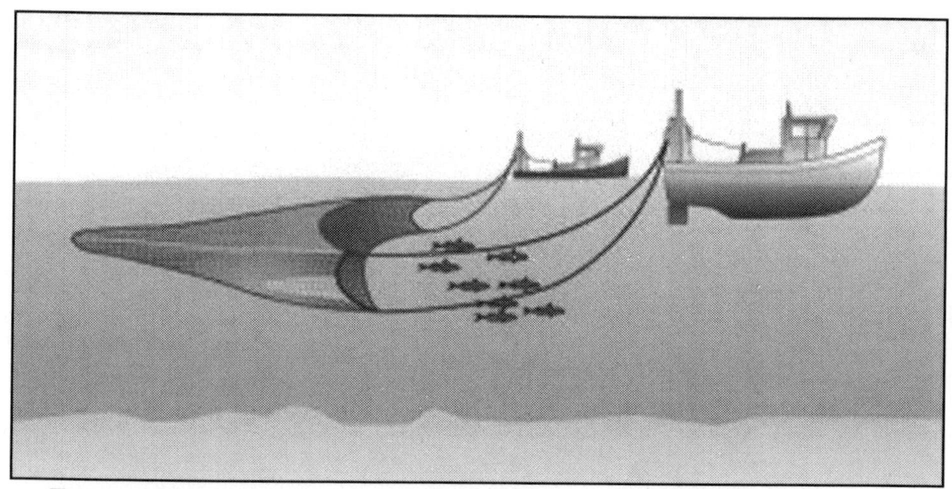

Funnel-shaped nets that are towed behind a fishing vessel through the water column at a predetermined depth, or along the seafloor. As the net travels through the water, fish are engulfed by the open mouth of the net and accumulate in the closed-off rear, called the cod end.

Angled panels at either side of the mouth, called otter boards, keep each end spread apart while the net is being pulled. Floats along the top and weights or a hydrodynamic depressor along the bottom ensure that the mouth remains vertically separated. There are two major subcategories of trawling: mid-water or pelagic trawling, and bottom or benthic trawling.

- Bottom trawling drags the trawl net directly along the ocean bottom. This direct contact results in a high amount of damage to benthic habitats, which are often slow to recover, or fail to recover at all.
- Mid-water trawling often employs larger nets, however, mid-water trawls do not typically come into contact with the bottom this technique causes less damage to marine habitats.

Both methods result in high rates of bycatch; particularly bottom trawls, which are notorious for harvesting large amounts of unwanted species including sea turtles and marine mammals. To combat this, the mesh-size of trawl nets used in some fisheries is regulated to permit smaller, undesired species and juvenile target species to evade capture by slipping through the netting. Devices to facilitate the escape of turtles, seals, and sea lions have also been developed.

These include metal grates placed towards the rear of the net and angled towards an escape portal cut into the top or bottom. The grate allows smaller fish and crustaceans to pass through and into the cod end, while turtles or marine mammals are directed through the portal. Trawling is one of the most widely used commercial fishing methods and is practiced around the world in many different fisheries. It is one of the methods used to catch Atlantic mackerel, anchovy, and sardine which are among the species listed on the KidSafe "Best Choices" seafood list. It is also used to catch certain varieties of flounder, orange roughy, Chilean Seabass, and other species.

GILLNET

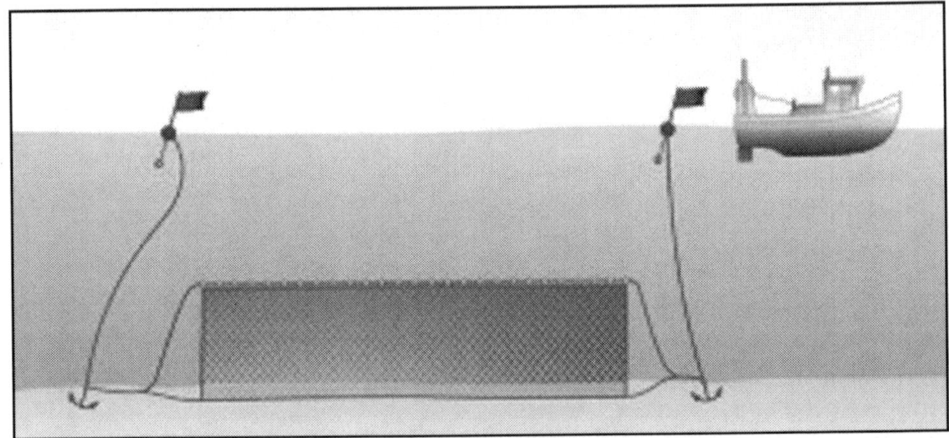

Gillnets are long strips or walls of netting that ensnare fish by their gills once they swim into the mesh. The nets are held in a vertical position by a series of floats running along the top and weights along the bottom. Gillnets may be set in a static position with the bottom running directly along the seafloor or suspended in the middle of the water column and attached to the seafloor with anchors.

Another type of gillnet, also called a driftnet, is set near the surface and is allowed to drift freely. Although a degree of selectivity is provided by the size of the mesh used, which depends on the fish species being targeted, this fishing technique is associated with high levels of bycatch.

Driftnets have been responsible for catching whales, dolphins, seals, sea turtles, and seabirds, as well as undesired fish species. They may also be lost or abandoned and become "ghost nets" that can drift for months, continuing to ensnare marine life. Due to these bycatch hazards the use of driftnets has been banned in some regions, and a regulation passed by the United Nations in 1991 limits the size of driftnets that can be used to 2. 5 kilometers in length. Before the U. N. ban, some driftnets were up to 60 kilometers long and 30 meters deep. Alaska salmon, which is included in the KidSafe "Best Choices" seafood list, is among the species targeted by stationary and drifting gillnets. The nets used, however, are much smaller than the driftnets employed in other fisheries and bycatch is kept to a minimum as schools of salmon are specifically targeted.

TROLL

Trolling is a technique that involves a fishing vessel towing a series of lines set at different depths, depending on location of the fish being targeted. When a fish takes the bait or artificial lure, the line is reeled in and the fish brought onboard. Trolling lines do not cause much damage to marine environments, as they typically do not come into contact with the bottom. This technique also has a low bycatch rate, and any accidental catches can be released after capture. Trolling is one of the methods used to harvest wild Alaskan salmon, which is among the species on the KidSafe "Best Choices" seafood list.

LONGLINE

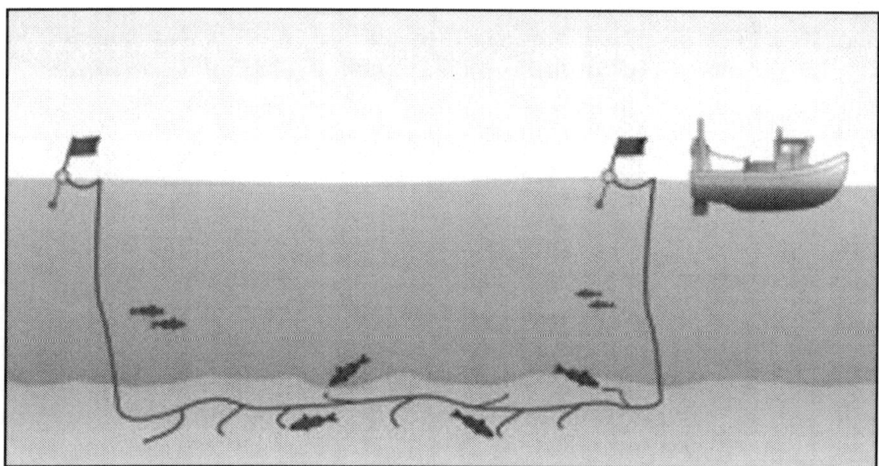

As the name suggests, a longline is composed of a mainline which may be greater than 100 kilometers in length with smaller lines suspended from it. A single long line may have well over 2000 of these smaller lines, each of which has a baited hook. As the fishing vessel moves forward, the mainline is unspooled from the sternand the baited branching lines are attached. The longline is then allowed to drift freely near the surface, or set on the bottom in some fisheries, for a pre-determined amount of time. After it has been allowed to "soak", the longline is recovered via buoys attached to the mainline and hauled back onboard. This method of fishing causes minimal habitat damage, however, is associated with high levels of bycatch. Because of the close proximity to the surface of many longlines, seabirds and turtles are common accidental captures. Swordfish and bluefin tuna are both species targeted by longlining, and should both be avoided.

HOOK AND LINE

Hook and line fishing involves lowering lines with baited hooks directly from the fishing vessel. When a fish takes the bait the line is retrieved either manually or with a mechanical device and the fish is brought onboard. This method causes minimal bycatch and habitat destruction.

HARPOON

Harpoons are used to catch large fish swimming near the surface; swordfish and bluefin tuna are two species captured with this method. Once the fish is spotted and within range, a harpoon is thrown or thrust into the fish. A barbed point on the harpoon prevents it from being pulled out as the fish tries to swim away. Vessels fitted for harpoon fishing are equipped with an elevated superstructure to give a wider field of vision for the spotter and with a pulpit in the bow to facilitate the harpooner getting within range of their quarry. This method is highly selective and therefore does not result in bycatch, however, species like swordfish and bluefin tuna are high in mercury, and should accordingly be avoided.

POTS AND TRAPS

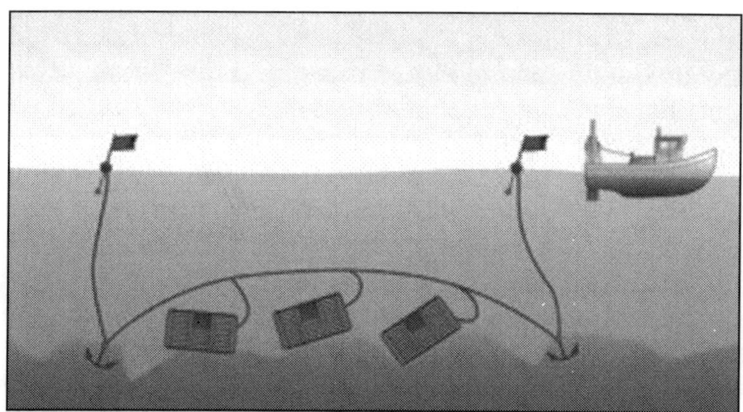

There are many different pot and trap designs, depending on the species being targeted. They are constructed of either a metal or wood frame with wire mesh or fabric netting, and have entrances that allow fish and/or crabs, lobsters, and other species in but hinder or prevent their escape. Bait is lowered to the bottom of the trap where they are allowed to sit or "soak". After a pre-determined length of time they are located, often by an attached buoy, and pulled back onboard the fishing vessel, sometimes by hand but mostly with a mechanized winch. As the traps and pots sit on the bottom, they cause little damage to the seafloor. Bycatch associated with traps and pots is also limited.

FISH FARMS

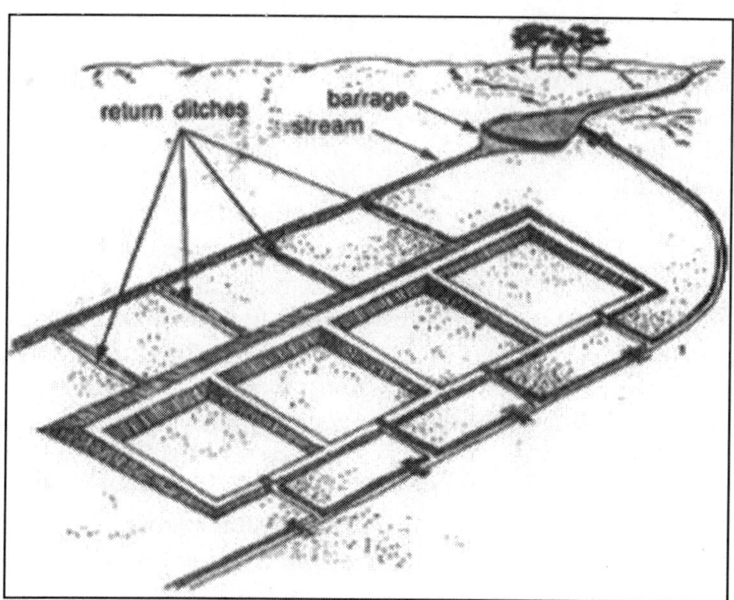

U. N. Food & Agriculture Organization

Various types of harvest operations raise groups of target species. There are four major types of fish farm: Open pens, ponds, raceways, and recirculating closed systems. Open pens are placed in a near-shore ocean location or in a lake and contain fish in a large net or wire cage that is anchored to the bottom and suspended from floats at the surface in direct contact with their surrounding ecosystems. Wastewater is passed freely to the surrounding area, and disease and parasites may be passed to fish swimming in near the pen.

Escapees may also compete and interbreed with wild fish, which can degrade the health of these natural populations. Ponds are placed in tidal or inland areas and may be salt or freshwater. If the wastewater they generate is not treated, it can contaminate the surrounding ecosystem(s), and the construction of pond farms has also been the cause of habitat destruction. Raceway farms divert water from a natural source, such as a stream, through channels where the fish are contained. The water flows through these channels and is then directed back into its original source, which can be polluted if the water coming from the farm is not properly treated. There is also the risk of fish escaping from the channels, and having harmful effects on wild fish populations. Recirculating closed systems treat and recycle the wastewater they generate. They also leave no chance of escape, which makes them completely cut off from the surrounding environment. Most farmed arctic char, included on the KidSafe "Best Choices" seafood list, is raised in recirculating closed systems. Rainbow trout, another fish from the list, are also sustainably farmed with minimal environmental impact.

ANGLING

Angling is a method of fishing by means of an "angle" (hook). The hook is usually attached to a line, and is sometimes weighed down by a sinker so it sinks in the water. This is the classic "hook, line and sinker" arrangement, used in angling since prehistoric times. The hook is usually baited with lures or bait fish. Additional arrangements include the use of a fishing rod, which can be fitted with a reel, and functions as a delivery mechanism forcasting the line. Other delivery methods for projecting the line include fishing kites and cannons, kontiki rafts and remote controlled devices. Floats can also be used to help set the line or function as bite indicators. The hook can be dressed with lures or bait. Angling is the principal method of sport fishing, but commercial fisheries also use angling methods involving multiple hooks, such as longlining or commercial trolling.

LINE FISHING

Line fishing is fishing with a fishing line. A fishing line is any cord made for fishing. Important parameters of a fishing line are its length, material, and weight (thicker, sturdier lines are more visible to fish). Factors that may determine what line an angler chooses for a given fishing environment include breaking strength, knot strength, UV resistance, castability, limpness, stretch, abrasion resistance, and visibility.

Modern fishing lines are usually made from artificial substances. The most common type is monofilament, made of a single strand. There are also braided fishing lines and thermally fused superlines.

- Droplining - a dropline consists of a long fishing line set vertically down into the water, with a series of baited hooks Droplines have aweight at the bottom and a float at the top. They are not usually as long as longlines and have fewer hooks.
- Handlining - is fishing with a single fishing line, baited with lures or bait fish, which is held in the hands. Handlining can be done from boats or from the shore. It is used mainly to catch groundfish and squid, but smaller pelagic fish can also be caught.
- Jiggerpole -
- Jigging - is the practice of fishing with a jig, a type of fishing lure. A jig consists of a lead sinker with a hook molded into it and usually covered by a soft body to attract fish. Jigs are intended to create a jerky, vertical motion, as opposed to spinnerbaits which move through the water horizontally.
- Longlining - is a commercial technique that uses a long heavy fishing line with a series of hundreds or even thousands of baited hooks hanging from the main line by means of branch lines called "snoods". Longlines are usually operated from specialised boats called longliners. They use a special winch to haul in the line, and can operate in deeper waters targeting pelagic species such as swordfish, tuna, halibut and sablefish.
- Slabbing: is a bass fishing technique, that involves repetitively lifting and dropping a flat lure, usually made of 1 to 2. 5 oz of lead painted to look like a baitfish (or heavy slabs of metal), through a school of actively feeding fish that the angler has located on afishfinder. Used on white and striped bass in the reservoirs of the southern USA.
- Trolling - is fishing with one or more baited lines which are drawn through the water. This may be done by pulling the line behind a slow moving boat, or by slowly winding the line in when fishing from the land. Trolling is used to catch pelagic fish such as mackereland kingfish.
- Trotlining - a trotline is like a dropline, except that a dropline has a series of hooks suspended vertically in the water, while a trotline has a series of hooks suspended horizontally in the water. Trotlines can be physically set in many ways, such as tying each end to

something fixed, and adjusting the set of the rest of the line with weights and floats. They are used for catching crabs or fish, such as catfish, particularly across rivers.

ANGLING WITH A ROD

Fishing rods give more control of the fishing line. The rod is usually fitted with a fishing reel which functions as a mechanism for storing, retrieving and paying out the line. Floats may also be used, and can function as bite indicators. The hook can be dressed with lures orbait.

- Bank fishing - fishing from river banks and similar shorelines. Bank fishing is usually performed with a fishing rod and reel, although nets, traps, and spears can also be used. People who fish from a boat can sometimes access more areas in prime locations with greater ease than bank fishermen. However many people don't own boats and find fishing from the bank has its own advantages. Bank fishing has its own requirements, and many things come into play for success, such as local knowledge, water depth, bank structure, location, time of day, and the type of bait and lures.
- Casting - the act of throwing the fishing line out over the water using a flexible fishing rod. The usual technique is for the angler to quickly flick the rod from behind towards the water. Casting is also a sport adjunct to fishing, much as shooting is to hunting. The sport is supervised by the International Casting Sport Federation, which sponsors tournaments and recognizes world records for accuracy and distance. Some variations of the technique exist, such as Surf fishing, the Reach cast, and Spey casting.
- Float tubes - small doughnut-shaped boats with an underwater seat in the "hole". Float tubes are used for fly fishing and enable the angler to reach deeper water without splashing and disturbing stillwater fish.
- Fly fishing - the use of artificial flies as lures. These are cast with specially constructed fly rods and fly lines. The fly line (today, almost always coated with plastic) is heavy enough cast in order to send the fly to the target. Artificial flies vary dramatically in size, weight and colour. Fly fishing is a distinct and ancient angling method, most renowned as a method for catching trout and salmon, but employed today for a wide variety of species including pike, bass, panfish, and carp, as well as marine species, such as redfish,snook, tarpon, bonefish and striped bass. There is a growing population of anglers whose aim is to catch as many different species as possible with the fly.
- Tenkara fishing - Tenkara is a form of fly fishing that originated in Japan over 200 years ago. It was originally done with a bamboo pole between 12' and 20' with the line tied directly on the tip of the rod

requiring no reel. Modern tenkara rods are usually made of graphite and are telescopic. Unlike western style fly fishing tenkara uses either a tapered line or a level line and forgoes the PVC coated fly fishing line. Typical target species include trout and char but most smaller freshwater species can be caught by this method.

- Rock fishing - fishing from rocky outcrops into the sea. It is a popular pastime in Australia and New Zealand. It can be a dangerous pastime and claims many lives each year.
- Surfcasting - fishing from a shoreline using a rod to cast into the surf. With few exceptions, surf fishing is done in saltwater, often from abeach. The basic idea of most surfcasting is to cast a bait or lure as far out into the water as is necessary to reach the target fish from the shore. This may or may not require long casting distances and muscular techniques. Basic surf fishing can be done with a surfcasting rodbetween seven and twelve feet long, with an extended butt section, equipped with an appropriate spinning or conventional casting reel. Dedicated surfcasters usually possess an array of terminal and other tackle, with rods and reels of different lengths and actions, and lures and baits of different weights and capabilities. Depending on fishing conditions and the fish they are targeting, such surfcasters tailor bait and terminal tackle to rod and reel and the size and species of the fish. Reels and other equipment need to be constructed so they resist the corrosive and abrasive effects of salt and sand.

OTHER ANGLING

- Bottom fishing - is fishing the bottom of a body of water. In the United Kingdom it is called "ledgering". A common rig for fishing on the bottom is a weight tied to the end of the line, with a hook about an inch up line from the weight. The method can be used both with hand lines and rods. There are fishing rods specialized for bottom fishing, called "donkas". The weight is used to cast or throw the line an appropriate distance. Bottom fishing can be done both from boats and from the land. It targets groundfishsuch as sucker fish, bream, catfish, and crappie.
- Ice fishing - is the practice of catching fish with lines and hooks through an opening in the ice on a frozen body of water. It is practised by hunter-gatherers such as the Inuitand by anglers in other cold or continental climates.
- Kayak fishing - has a long history, and has gained popularity in recent times. Many of the techniques used are the same as those used on other fishing boats, apart from difference is in the set-up, how each piece of equipment is fitted to the kayak, and how each activity is carried out on such a small craft.

- Kite fishing is said to have been invented in China. It was, and still is, used by the people of New Guinea and other Pacific Islands - either by cultural diffusion from China or independent invention. Kites can provide the boatless fishermen access to waters that would otherwise be available only to boats. Similarly, for boat owners, kites provide a way to fish in areas where it is not safe to navigate such as shallows or coral reefs where fish may be plentiful. Kites can also be used for trolling a lure through the water. Suitable kites may be of very simple construction. Those of Tobi Island are a large leaf stiffened by the ribs of the fronds of the coconut palm. The fishing line may be made from coconut fibre and the lure made from spiders webs. Modern kitefishing is popular in New Zealand, where large delta kites of synthetic materials are used to fish from beaches, taking a line and hooks far out past the breakers. Kite fishing is also emerging in Melbourne where sled kites are becoming popular, both off beaches and off boats and in freshwater areas. The disabled community are increasingly using the kites for fishing as they allow mobility impaired people to cast the bait further out than they would otherwise be able to.
- Boat anglers - Fishing is usually done either from a boat or from a shoreline or river bank. When fishing from a boat, pretty much any fishing technique can be used, from nets to fish traps, but some form of angling is by far the most common. Compared to fishing from the land, fishing from a boat allows more access to different fishing grounds and different species of fish. Some tackle is specialised for boat anglers, such as sea rods.
- Remote control fishing - Fishing can also be done using a remote controlled boat. This type of fishing is commonly referred to as RC fishing. The boat is usually one to three feet long and runs on a small DC battery. A radio transmitter controls the boat. The fisherman connects the fishing line/bait to the boat; drives it; navigating the water by manipulating the remote controller. The technique is growing in popularity.

LAWS AND REGULATIONS

Laws and regulations managing angling vary greatly, often regionally, within countries. These commonly include permits (licences), closed periods (seasons) where specific species are unavailable for harvest, restrictions on gear types, and quotas.

Laws generally prohibit catching fish with hooks other than in the mouth (foul hooking, “snagging” or “jagging”) or the use of nets other than as an aid in landing a captured fish. Some species, such as bait fish, may be taken with nets, and a few for food. Sometimes, (non-sport) fish are considered of lesser value and it may be permissible to take them by methods like snagging, bow and arrow, or spear. None of these techniques fall under the definition of angling since they do not rely upon the use of a hook and line.

FISHING SEASONS

Fishing seasons are set by countries or localities to indicate what kinds of fish may be caught during sport fishing (also known as angling) for a certain period of time. Fishing seasons are enforced to maintain ecological balance and to protect species of fish during theirspawning period during which they are easier to catch.

SLOT LIMITS

Slot limits are put in action to help protect certain fish in given area. They generally require anglers to release captured fish if they fall within a given size range, allowing anglers to keep only smaller or larger fish. Slot Limits vary from lake to lake depending on what local officials believe would produce the best outcome for managing fish populations.

CATCH AND RELEASE

Although most anglers keep their catch for consumption, catch and release fishing is increasingly practised, especially by fly anglers. The general principle is that releasing fish allows them to survive, thus avoiding unintended depletion of the population.

For species such as marlin and muskellunge but, also, among many bass anglers, there is a cultural taboo against killing bass for food. In many parts of the world, size limits apply to certain species, meaning fish below a certain size must, by law, be released. It is generally believed that larger fish have a greater breeding potential. Some fisheries have a slot limit that allows the taking of smaller and larger fish, but requiring that intermediate sized fish be released. It is generally accepted that this management approach will help the fishery create a number of large, trophy-sized fish. In smaller fisheries that are heavily fished, catch and release is the only way to ensure that catchable fish will be available from year to year.

The practice of catch and release is criticised by some who consider it unethical to inflict pain upon a fish for purposes of sport. Some of those who object to releasing fish do not object to killing fish for food. Adherents of catch and release dispute this charge, pointing out that fish commonly feed on hard and spiky prey items, and as such can be expected to have tough mouths, and also that some fish will re-take a lure they have just been hooked on, a behaviour that is unlikely if hooking were painful. Opponents of catch and release fishing would find it preferable to ban or to severely restrict angling. On the other hand, proponents state that catch-and-release is necessary for many fisheries to remain sustainable, is a practice that that generally has high survival rates, and consider the banning of angling as not reasonable or necessary.

Fig. Removing the hook from a Bonito.

In some jurisdictions, in the Canadian province of Manitoba, for example, catch and release is mandatory for some species such as brook trout. Many of the jurisdictions which mandate the live release of sport fish also require the use of artificial lures and barbless hooks to minimise the chance of injury to fish. Mandatory catch and release also exists in the Republic of Ireland where it was introduced as a conservation measure to prevent the decline of Atlantic salmon stocks on some rivers. In Switzerland, catch and release fishing is considered inhumane and was banned in September 2008.

Barbless hooks, which can be created from a standard hook by removing the barb with pliers or can be bought, are sometimes resisted by anglers because they believe that increased escapement results. Barbless hooks reduce handling time, thereby increasing survival. Concentrating on keeping the line taut while fighting fish, using recurved point or "triple grip" style hooks on lures, and equipping lures that do not have them with split rings can significantly reduce escapement.

CAPACITY FOR PAIN

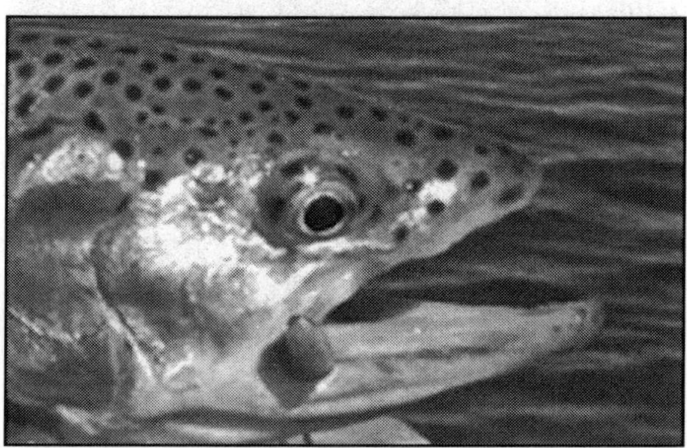

Fig. Rainbow trout.

Animal protection advocates have raised concerns about the suffering of fish caused by angling. In light of recent research, some countries, like Germany, have banned specific types of fishing and the British RSPCA now formally prosecutes individuals who are cruel to fish.

Experiments done by William Tavolga provide evidence that fish have pain and fear responses. For instance, in Tavolga's experiments,toadfish grunted when electrically shocked and over time they came to grunt at the mere sight of an electrode. Additional tests conducted at both the University of Edinburgh and the Roslin Institute, in which bee venom and acetic acid was injected into the lips of rainbow trout, resulted in fish rubbing their lips along the sides and floors of their tanks, which the researchers believe was an effort to relieve themselves of pain. One researcher argues about the definition of pain used in the studies.

In 2003, Scottish scientists at the University of Edinburgh performing research on rainbow trout concluded that fish exhibit behaviours often associated with pain, and the brains of fish fire neurons in the same way human brains do when experiencing pain. James D. Rose of the University of Wyoming critiqued the study, claiming it was flawed, mainly since it did not provide proof that fish possess "conscious awareness, particularly a kind of awareness that is meaningfully like ours". Rose argues that since the fish brain is rather different from ours, fish are not conscious, whence reactions similar to human reactions to pain instead have other causes. Rose had published his own opinion a year earlier arguing that fish cannot feel pain as they lack the appropriate neo-cortex in the brain. However, animal behaviourist Temple Grandin argues that fish could still have consciousness without a neo-cortex because "different species can use different brain structures and systems to handle the same functions. " The position that Rose takes also fails to address unresolved empirical and philosophical considerations concerning pain, as raised by principles ofepistemology, solipsism, existentialism, and comparative ethology. Until such problems are far more fundamentally resolved, there are strong arguments for refraining from causing the appearance of pain or behaviour consistent with pain, insofar as such things might be reasonably avoidable. However, in 2012, a group of researchers led by Rose reviewed the literature, and concluded again that fish are not conscious and therefore do not feel pain.

TOURNAMENTS AND DERBIES

Fig. Angling at Shihtiping in Taiwan.

Sometimes considered within the broad category of angling is where contestants compete for prizes based on the total length or weight of a fish, usually of a pre-determined species, caught within a specified time (fishing tournaments). Such contests have evolved from local fishing contests into large competitive circuits, where professional anglers are supported by commercial endorsements.

Professional anglers are not engaged in commercial fishing, even though they gain an economic reward. Similar competitive fishing exists at the amateur level with fishing derbies. In general, derbies are distinguished from tournaments; derbies normally require fish to be killed. Tournaments normally deduct points if fish can not be released alive.

MOTIVATION

A ten-year-long survey of US fishing club members, completed in 1997, indicated that motivations for recreational angling have shifted from relaxation, an outdoor experience and the experience of the catch, to the importance of family recreation. Anglers with higher family incomes fished more frequently and were less concerned about obtaining fish as food. A German study indicated that satisfaction derived from angling was not dependent on the actual catch, but depended more on the angler's expectations of the experience. A 2006 study by the Louisiana Department of Wildlife and Fisheries tracked the motivations of anglers on the Red River. Among the most often stated responses were the fun of catching fish, the experience, to catch a lot of fish or a very large fish, for challenge, and adventure. Use as food was not investigated as a motive.

TRAPPING

Fig. Fishermen with traditional fish traps,Hà Tây, Vietnam.

Fig. A typical wooden fish wheel.

Fig. Lobster pots on the beach at Beer, Devon.

Traps are culturally almost universal and seem to have been independently invented many times. There are essentially two types of trap, a permanent or semi-permanent structure placed in a river or tidal area and pot-traps that are baited to attract prey and periodically lifted.

Artisanal Techniques

- Dam fishing - An artisanal technique called dam fishing is used by the Baka pygmies. This involves the construction of a temporary dam resulting in a drop in the water levels downstream—allowing fish to be easily collected.
- Basket weir fish traps - were widely used in ancient times. They are shown in medieval illustrations and surviving examples have been found. Basket weirs are about 2 m long and comprise two wicker cones, one inside the other—easy to get into and hard to get out.
- Fishing weir - In medieval Europe, large fishing weir structures were constructed from wood posts and wattle fences. 'V' shaped structures in rivers could be as long as 60 metres and worked by directing fish towards fish traps or nets. Such fish traps were evidently controversial in medieval England. The Magna Carta includes a clause requiring that they be removed: "All fish-weirs shall be removed from the Thames, the Medway, and throughout the whole of England, except on the sea coast".
- Fish wheels - operate alongside streams, much as a water-powered mill wheel. A wheel complete with baskets and paddles is attached to a floating dock. The wheel rotates due to the current of the stream. The baskets on the wheel capture fish travelling upstream and transfer them into a holding tank. When the holding tank is full, the fish are removed.
- Lobster traps - also called lobster pots, are traps used to catch lobsters. They resemble fish traps, yet are usually smaller and

consist of several sections. Lobster traps are also used to catch other crustaceans, such as crabs and crayfish. They can be constructed in various shapes, but the design strategy is to make the entry into the trap much easier than exit. The pots are baited and lowered into the water and checked frequently. Historically lobster pots were constructed with wood or metal. Today most traps are made from checkered wire and mesh. It is common for the trap to be weighted down with bricks. A bait bag is hung in the middle of the trap. In theory the lobster walks up the mesh and then falls into the wire trap. Bait varies from captain to captain but it is common to use herring. In commercial lobstering five to ten of these traps will be connected with line. A buoy marks each end of the string of pots. Two buoys are important to make retrieval easier and so captains don't set their traps over each other. Each buoy is painted differently so the various captains can identify their traps.

ANIMALS

- Cooperative human-dolphin fisheries date back to the ancient Roman author and natural philosopher Pliny the Elder. A modern human-dolphin fishery still takes place in Laguna, Santa Catarina, Brazil. Here, dolphins drive fish towards fishermen waiting along the shore and give them a signal when they can cast their nets. The dolphins then feed off the fish that manage to escape the nets.
- Cormorant fishing - In China and Japan, the practice of cormorant fishing is thought to date back some 1300 years. Fishermen use the natural fish-hunting instincts of the cormorants to catch fish, but a metal ring placed round the bird's neck prevents large, valuable fish from being swallowed. The fish are instead collected by the fisherman.
- Frigatebirds fishing - The people of Nauru used trained frigatebirds to fish on reefs.
- Portuguese Water Dogs - Dating from the 16th century in Portugal, Portuguese Water Dogs were used by fishermen to send messages between boats, to retrieve fish and articles from the water, and to guard the fishing boats. Labrador Retrievers have been used by fishermen to assist in bringing nets to shore; the dog would grab the floating corks on the ends of the nets and pull them to shore.
- Remora fishing - The practice of tethering a remora, a sucking fish, to a fishing line and using the remora to capture sea turtles probably originated in the Indian Ocean. The earliest surviving records of the practice are Peter Martyr d'Anghera's 1511 accounts of the second voyage of Columbus to the New World (1494). However, these

accounts are probably apocryphal, and based on earlier, no longer extant accounts from the Indian Ocean region.

OTHER TECHNIQUES

Fig. Scientists carrying out a population and species survey using electrofishing equipment.

Christ catches fish using a miracle technique

ARTISANAL TECHNIQUES

Basnig -

- Electrofishing - is another recently developed technique, primarily used in freshwater by fisheries scientists. Electrofishing uses electricity to stun fish so they can be caught. It is commonly used in scientific surveys, sampling fish populations for abundance, density, and species composition. When performed correctly, electrofishing results in no permanent harm to fish, which return to their natural state a few minutes after being stunned.
- Fish aggregating devices - are man-made objects used to attract pelagic fish such as marlin, tuna and mahi-mahi (dolphin fish). They usually consist of buoys or floats tethered to the ocean floor with concrete blocks.
- Dredging - There are types of dredges used for collecting scallops, oysters or sea cucumbers from the seabed. They have the form of a scoop made of chain mesh and they are towed by a fishing boat. Dredging can be destructive to the seabed, because the marine life

is unable to survive the weight of the dredge. It is extremely detrimental to coral beds since they take centuries to rebuild themselves. Unmonitored dredging can be compared to unmonitored forest clearing, where it can wipe out ecosystems. Nowadays, this method of fishing is often replaced by mariculture or by scuba diving.

- Fish finders - are electronic sonar devices which indicate the presence of fish and fish schools. They are widely used by recreational fishermen. Commercially, they are used with other electronic locating and positioning devices.
- Fishing light attractors - use lights attached (above or underwater) to some structure to attract fish and bait fish. Fishing light attractor are operated every night. After a while, fish discover the increased concentration of bait surrounding the light. Once located, the fish return regularly, and can be harvested.

Flossing -

- Harvesting machines - have recently been developed for commercial fishing. Harvesting machines use pumps to pump fish out of the sea. Dredges have also been mechanized so that they directly transfer mollusks to the surface as are dredged.
- Payaos - a type of fish aggregating device used in Southeast Asia, particularly in the Philippines. Payaos were traditionally bamboo rafts for handline fishing before World War II, but modern steel payaos use fish lights and fish location sonar to increase yields. While payaos fishing is sustainable on a small scale, the large scale, modern applications have been linked to adverse impacts on fish stocks.
- Shrimp baiting - is a method used by recreational fisherman for of catching shrimp. It uses a cast net, bait and long poles. The poles are used to mark a specific location and then bait is thrown in the water near the pole. After several minutes the cast net is thrown as close to the bait as possible and shrimp are caught in the net. In the 1980s the sport became popular in the south eastern coastal states of the USA.

DESTRUCTIVE TECHNIQUES

Destructive fishing practices are practices that easily result in irreversible damage to aquatic habitats and ecosystems. Many fishing techniques can be destructive if used inappropriately, but some practices are particularly likely to result in irreversible damage. These practices are mostly, though not always, illegal. Where they are illegal, they are often inadequately enforced.

Some examples are:

- Explosives - Dynamite or blast fishing, is done easily and cheaply with dynamite or homemade bombs made from locally available materials. Fish are killed by the shock from the blast and are then skimmed from the surface or collected from the bottom. The explosions indiscriminately kill large numbers of fish and other marine organisms in the vicinity and can damage or destroy the physical environment. Explosions are particularly harmful to coral reefs. Blast fishing is also illegal in many waterways around the world.

BOTTOM TRAWLING

- Cyanide fishing - Cyanides are used to capture live fish near coral reefs for the aquarium and seafood market. This illegal fishing occurs mainly in or near the Philippines,Indonesia, and the Caribbean to supply the 2 million marine aquarium owners in the world. Many fish caught in this fashion die either immediately or in shipping. Those that survive often die from shock or from massive digestive damage. The high concentrations of cyanide on reefs harvested in this fashion damages the coral polyps and has also resulted in cases of cyanide poisoning among local fishermen and their families.

FISH TOXINS

- Muroami - is a destructive artisan fishing method employed on coral reefs in Southeast Asia. An encircling net is used with pounding devices, such as large stones fitted on ropes that are pounded onto the coral reefs. They can also consist of large heavy blocks of cement suspended above the sea by a crane fitted to the vessel. The pounding devices are repeatedly lowered into the area encircled by the net, smashing the coral into small fragments in order to scare the fish out of their coral refuges. The “crushing” effect on the coral heads has been described as having long-lasting and practically totally destructive effects.

HISTORY

Ancient remains of spears, hooks and fishnet have been found in ruins of the Stone Age. The people of the early civilization drew pictures of nets and fishing lines in their arts (Parker 2002). Early hooks were made from the upper bills of eagles and from bones, shells, horns and plant thorns. Spears were tipped with the same materials, or sometimes with flints.

Lines and nets were made from leaves, plant stalk and cocoon silk. Ancient fishing nets were rough in design and material but they were amazingly, as if some now use (Parker 2002). Literature on the indigenous fishing practices is very scanty. Baines (1992) documented traditional fisheries in the Solomon Islands.

Use of the herbal fish poisons in catching fishes from fresh water and sea documented from New Caledonia (Dahl 1985). John (1998) documented fishing techniques and overall life style of the Mukkuvar fishing Community of Kanyakumari district of Tamil Nadu, India. Tribal people using various plants for medicinal and various purposes extends the use notion for herbal fish stupefying plants. Use of the fish poisons is very old practice in the history of human kind. In 1212, King Frederick II prohibited the use of certain plant piscicides, and by the 15th century similar laws had been decreed in other European countries as well (Wilhelm 1974). All over the globe, indigenous people use various fish poisons to kill the fishes, documented in America (Jeremy 2002) and among Tarahumara Indian (Gajdusek 1954).

SPEARFISHING

Spearfishing is an ancient method of fishing that has been used throughout the world for millennia. Early civilizations were familiar with the custom of spearing fish from rivers and streams using sharpened sticks. Today modern spearfishing makes use of elastic powered spearguns and slings, or compressed gas pneumatic powered spearguns, to strike the hunted fish. Specialised techniques and equipment have been developed for various types of aquatic environments and target fish.

Spearfishing may be done using free-diving, snorkelling, or scuba diving techniques. Spearfishing while using scuba equipment is illegal in some countries. The use of mechanically powered spearguns is also outlawed in some countries and jurisdictions. Spearfishing is highly selective, normally uses no bait and has no by-catch.

HISTORY

Spearfishing with barbed poles (harpoons) was widespread in palaeolithic times. Cosquer cave in Southern France contains cave art over 16,000 years old, including drawings of seals which appear to have been harpooned. There are references to fishing with spears in ancient literature; though, in most cases, the descriptions do not go into detail. An early example from the Bible is in Job 41:7: Canst thou fill his [Leviathan] skin with barbed irons? or his head with fish spears?.

The Greek historian Polybius (ca 203 BC–120 BC), in his Histories, describes hunting for swordfish by using a harpoon with a barbed and detachable head. Greek author Oppian of Corycus wrote a major treatise on sea fishing, the Halieulica or Halieutika, composed between 177 and 180. This is the earliest such work to have survived intact. Oppian describes various means of fishing including the use of spears and tridents.

In a parody of fishing, a type of gladiator called retiarius carried a trident and a casting-net. He fought the murmillo, who carried a short sword and a helmet with the image of a fish on the front. Copper harpoons were known to the seafaring Harappans well into antiquity. Early hunters in India include the Mincopie people, aboriginal inhabitants of India's Andaman and Nicobar islands, who have used harpoons with long cords for fishing since early times.

TRADITIONAL

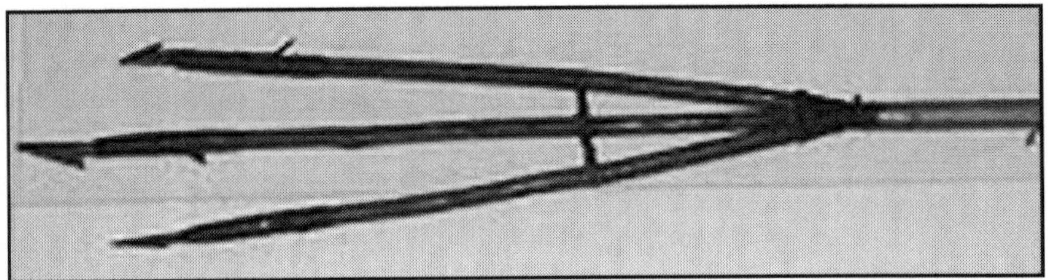

Fig. Head of an arrow used for fishing, from Guyana.

Spear fishing is an ancient method of fishing and may be conducted with an ordinary spear or a specialised variant such as an eel spear or the trident. A small trident type spear with a long handle is used in the American South and Midwest for gigging bullfrogs with a bright light at night, or for gigging carp and other fish in the shallows. Traditional spear fishing is restricted to shallow waters, but the development of the speargun allows fishing in deeper waters. With practice, divers are able to hold their breath for up to four minutes and sometimes longer; of coursc, a diver with underwater breathing equipment can dive for much longer periods.

MODERN

In the 1920s, sport spearfishing using only watertight swimming goggles became popular on the Mediterranean coast of France and Italy. This led to development of the moderndiving mask, fins and snorkel. Modern scuba diving had its genesis in the systematic use of rebreathers by Italian sport spearfishers during the 1930s. This practice came to the attention of the Italian Navy, which developed its frogman unit, which affected World War II. By 1940 small groups of people in California, USA had been spearfishing for less than 10 years. Most used imported gear from Europe, while innovators Charlie Sturgill, Jack Prodanovich, and Wally Potts invented and built innovative equipment for California divers.

During the 1960s, attempts to have spearfishing recognised as an Olympic sport were unsuccessful. Instead, two organisations, the International Underwater Spearfishing Association (IUSA) and the International Bluewater Spearfishing Records Committee (IBSRC), list world record catches by species according to rules to ensure fair competition. Spearfishing is illegal in many bodies of water, and some locations only allow spearfishing during certain seasons.

Conservation

Spearfishing has been implicated in local disappearances of some species, including the Atlantic goliath grouper on the Caribbean island of Bonaire- the Nassau grouper in the barrier reef off the coast of Belize, the giant black sea bass in California, which have all been listed as endangered. Modern Spearfishing has shifted focus onto catching only what one needs and targeting sustainable fisheries. As gear evolved in the 1960s and 1970s spearfishermen typically viewed the ocean as an unlimited resource and often sold their catch.

This practise is now heavily frowned upon in prominent spearfishing nations for promoting unsustainable methods and encouraging taking more fish than is needed. In countries such as Australia and South Africa where the activity is regulated by state fisheries, spearfishing has been found to be the most environmentally friendly form of fishing due to being highly selective, having no by-catch, causing no habitat damage, nor creating pollution or harm to protected endangered species. In 2007, the Australian Bluewater Freediving Classic became the first spearfishing tournament to be accredited and was awarded 4 out of 5 stars based on environmental, social, safety and economic indicators.

Shore Diving

Fig. Spearfisherman hunting Yellowfin tuna in the Ryukyu Islands.

Shore diving is perhaps the most common form of spearfishing and simply involves entering and exiting the sea from beaches orheadlands and hunting around ocean structures, usually reef, but also rocks, kelp or sand. Usually shore divers hunt at depths of 5–25 metres (16–82 ft), depending on location. In some locations in the South Pacific, divers can experience drop-offs from 5 to 40 metres (16 to 131 ft) close to the shore line. Sharks and reef fish can be abundant in these locations. In subtropical areas, sharks may be less common, but other challenges face the shore diver, such as managing entry and exit in the presence of big waves. Headlands are favoured for entry because of their proximity to deeper water, but timing is important so the diver does not get pushed onto rocks by waves. Beach entry can be safer, but more difficult due the need to consistently dive through the waves until the surf line is crossed.

Shore dives produce mainly reef fish, but oceangoing pelagic fish are caught from shore dives too, and can be specifically targeted. Shore diving can be done with trigger-less spears such as pole spears or Hawaiian slings, but more commonly triggered devices such as spearguns. Speargun setups to catch and store fish include speed rigs and fish stringers.

Boat Diving

Boats, ships, kayaks, or even jetski can be used to access offshore reefs or ocean structure. Man-made structures such as oil rigs and Fish Aggregating Devices (FADs) are also fished. Sometimes a boat is necessary to access a location that is close to shore, but inaccessible by land. Methods and gear used for boat diving are similar to shore diving or blue water hunting, depending on the target prey.

Boat diving is practised worldwide. Hot spots include Mozambique, the Three Kings islands of New Zealand (yellowtail), Gulf of Mexico oil rigs (cobia, grouper) and the Great Barrier Reef (wahoo, dogtooth tuna). The deepwater fishing grounds off Cape Point, (Cape Town, South Africa) have become popular with trophy hunting, freediving spearfishers in search of Yellowfin Tuna.

Blue Water Hunting

Blue water hunting involves diving in open ocean waters for pelagic species. It involves accessing usually very deep and clear water and chumming for large pelagic fish species such as marlin, tuna, wahoo, or giant trevally. Blue water hunting is often conducted in drifts; the boat driver drops divers and allow them to drift in the current for up to several kilometres before collecting them. Blue water hunters can go for hours without seeing any fish, and without any ocean structure or a visible bottom the divers can experiencesensory deprivation and have difficulty determining the size of a solitary fish. One technique to overcome this is to note the size of the fish's eye in relation to its body. Large specimens have a proportionally smaller eye.

The creation of the Australian Bluewater Freediving Classic in 1995 in northern New South Wales was a revolutionary way of creating interest and promotion of this format of conservative underwater hunting, and contributed to the formation of the International Bluewater Spearfishing Records Committee. The IBSRC formed in 1996, was the first dedicated organization worldwide, created by recognized world leaders in blue-water hunting, to record and regulate the capture of pelagic species by blue-water hunters. Notably, some blue water hunters use large multi-band wooden guns and make use of breakaway rigs to catch and subdue their prey. If the prey is large and still has fight left after being subdued, a second gun can provide a kill shot at a safe distance. This is acceptable to IBSRC and IUSA regulations as long as the spearo loads it himself in the water. Blue water hunting is conducted worldwide, but notable hot spots include Mozambique (dogtooth tuna, wahoo and giant turrum), South Africa (Yellowfin tuna, Spanish Mackerel, wahoo, marlin and giant turrum), Australia (dogtooth tuna, wahoo and Spanish Mackerel) and the South Pacific (dogtooth tuna). Tanzania has been removed as a notable hot spot as spearfishing is illegal according to the laws and regulations of both Tanzania and Zanzibar.

FRESHWATER HUNTING

Fig. A common carp shot with a band-powered speargun by a diver using snorkelling gear, Minnesota, USA.

Many US states allow spearfishing in lakes and rivers, but nearly all of them restrict divers to shooting only rough fish such as carp, gar,bullheads, suckers, etc. A few US states do allow the taking of certain gamefish such as sunfish, crappies, striped bass, catfish andwalleyes. Freshwater hunters typically have to deal with widely varying seasonal changes in water clarity due to flooding, algae blooms and lake turnover. Some especially hardy midwestern and north central SCUBA divers go spearfishing under the ice in the winter when water clarity is at its best. In the summer the majority of freshwater spearfishermen use snorkelling gear rather than SCUBA since many of the fish they pursue are in relatively shallow water. Carp shot by freshwater spear fishermen typically end up being used as fertilizer, bait for trappers, or are occasionally donated to zoos.

WITHOUT DIVING

Fig. Night spear fishing, Amazon basin, Peru.

Spearfishing with a hand held spear from land, shallow water or boat has been practised for thousands of years. The fisher must account for optical refraction at the water's surface, which makes fish appear higher in their line of sight than they are. By experience, the fisher learns to aim lower. Calm and shallow waters are favoured for spearing fish from above the surface, as water clarity is of utmost importance. Many people who grew up on farms in the midwest U. S. in the 1940s-'60s recall going spearing for carp with pitchforks when their fields flooded in the spring. Spearfishing in this manner has some similarities to bowfishing.

EQUIPMENT

This is a list of equipment commonly used in spearfishing. Not all of it is necessary and spearfishing is often practised with minimal gear.

Speargun

A speargun is an underwater fishing implement designed to fire a spear at fish. The most popular spearguns are powered by natural latex rubber bands, while pnuematic powered guns are also used, but less powerful.

Polespear

Pole spears, or hand spears, consist of a long shaft with point at one end and an elastic loop at the other for propulsion. They also come in a wide variety, from aluminum or titanium metal, to fiberglass or carbon fibre. Often they are screwed together from smaller pieces or able to be folded down for ease of transport. In 1951 Charlie Sturgill beat the competition (who were all using spearguns) with his own pole spear design. This proved that a simple pole spear could be more effective than a speargun.

Hawaiian Slings

Hawaiian slings consist of an elastic band attached to a tube, through which a spear is launched.

Wet Suit

Wetsuits designed specifically for spearfishing are often two-piece (jacket and high waisted pants or 'long-john' style pants with shoulder straps) and are black or are fully or partially camouflage. Camouflage patterns include, blue for open ocean, green or brown for reef hunting. Commonly they have a pad on the chest to aid in loading spearguns. Sometimes they have reinforced elbows or knees. Often they have an "open-cell" interior, which offers superior warmth because of its very close suction fit to the skin (consequently lubrication is required to put on an open-cell wetsuit). Nylon exteriors are generally used to protect the spearo wetsuit from reefs but more delicate uncovered "smooth skin" wetsuits are sometimes used when diving from boats, because they dry quickly.

Weight Belt or Weight Vest

These are used to compensate for wetsuit buoyancy and help the diver descend to depth. Rubber belts, which can be quickly released in an emergency, have proven to be a particular popular design for spearfishing worldwide, such that most spearfishing equipment manufacturers now offer them.

Fins

Fins for freedive spearfishing are much longer than those used in SCUBA to aid in fast ascent. Typically a closed foot design is used by freediving (snorkelling) spearos, usually worn with neo-prene socks, while open foot designs (which allow diving boots to be worn) are more popular with SCUBA divers.

Knife or Cutters

A knife is carried as a safety precaution in case the diver becomes tangled in a spearline or floatline. It can also be used as an iki jime or kill spike.

Iki Jime or Kill Spike

In lieu of a knife, a sharpened metal spike can be used to kill the fish quickly and humanely upon capture. This action reduces interest from sharks by stopping the fish from thrashing. Iki jime is a Japanese term and is a method traditionally used by Japanese fishermen. Killing the fish quickly is believed to improve the flavour of the flesh by limiting the buildup of adrenaline in the fish's muscles.

Buoy or Float

A buoy is usually tethered to the spearfisher's speargun or directly to the spear. A buoy helps to subdue large fish. It can also assist in storing fish. But is more importantly used as a safety device to warn boat drivers there is diver in the area - usually by being large, brightly colored and flying a dive flag (the red-white "international" flag in the USA or the blue & white "alpha" flag elsewhere in the world). A typical spearo dive float will be torpedo-shaped, orange or red in colour with a volume of between 7 and 36 litres and display a dive flag on a short mast. However, other designs, such as inflatable mini-dinghy, planche (box), Tommy Botha (big game) and body-boards are also used.

Floatline

A floatline connects the buoy to the speargun or to the weight-belt. Often made from braided polyester, they are also frequently made from mono-filament encased in an airtight plastic tube, or made from stretchable bungee cord.

Gloves

Gloves are valuable to spearfishermen who desire to maintain a sense of safety or access more dangerous areas, such as those between coral, that could otherwise not be reached without use of the hands. They also aid in loading the bands on rubber powered speargun and protect the spearfisher's hands from the teeth and spines of struggling fish. They are also used for warmth in colder climes.

Fish Stringer

Usually a length of cable/cord/string/monofilament terminated by a loop (and sometimes a swivel) at one end and a large stainless steel pin/spike at the other. The pin is typically 15–30 cm long, 4-8mm diameter, with a sharp point at one end, and with the cable threaded through a hole, usually in the middle. Can alternatively be a large, shaped loop of stainless steel. Used to store speared fish on. Usually attached to the dive float or around the waist of the diver. The pin can optionally be used as an iki jimespike, to dispatch speared fish.

Snorkel and Diving Mask

Spearfishing snorkels and diving masks are similar to those used for scuba diving, although the masks usually have two lenses and a lower internal volume.

Diver Down Flag

The "diver down" flag (also called a "dive flag") is a safety flag used on the water to indicate to other boats that there is a diver below. When in use, it signals to other vessels to keep clear, watch for divers in the water, and operate at a slow speed.

MANAGEMENT

Spearfishing is intensively managed throughout the world.

Australia allows only recreational spearfishing and generally only breath-hold free diving. State & territory governments impose numerous restrictions, demarcating Marine Protected Areas, Closed Areas, Protected Species, size/ bag limits and equipment. The body principally concerned with spearfishing is the Australian Underwater Federation, Australia's peak recreational diving body. The AUF's vision for spearfishing is "Safe, Sustainable, Selective, Spearfishing". The AUF provides membership, advocacy and organises competitions.

Norway has a relatively large ratio of coastline to population, and has one of the most liberal spearfishing rules in the northern hemisphere. Spearfishing with scuba gear is widespread among recreational divers. Restrictions in Norway are limited to anadrome species, like atlantic salmon, sea trout, and lobster. In Mexico a regular fishing permit allows spearfishing, but not electro-mechanical spearguns. Spearfishing with scuba gear is illegal and the use of power heads as well. Penalties are severe and include fines, confiscated gear and even imprisonment.

United States has different spearfishing regulations for each state. In Florida spearfishing is restricted to several hundred yards offshore in many areas and the usage of apowerhead is prohibited within state waters. Many types of fish are currently under heavy bag restrictions. In California only recreational spearfishing is allowed. California also imposes numerous restrictions, demarcating Marine protected areas, closed areas, protected species, size/bag limits and equipment.

In the UK, while spearfishing is not explicitly regulated, it is instead subject to both local (typically local bye-laws) and national-level legislation relating to permitted fish species and minimum size limits. For example, it is not permitted to spearfish in freshwater and the non-tidal reaches of rivers. Under recent EU guidelines, recreational spearfishing is now explicitly permitted in the EUs Atlantic waters.

FISH TRAP

A fish trap is a trap used for fishing. Fish traps may have the form of a fishing weir or a lobster trap. A typical trap might consist of a frame of thick steel wire in the shape of a heart, with chicken wire stretched around it. The mesh wraps around the frame and then tapers into the inside of the trap. When a fish swims inside through this opening, it cannot get out, as the chicken wire opening bends back into its original narrowness. In earlier times, traps were constructed of wood and fibre.

HISTORY

Fig. Eel Traps in England, 1899, by Myles Birket Foster.

Traps are culturally almost universal and seem to have been independently invented many times. There are essentially two types of trap, a permanent or semi-permanent structure placed in a river or tidal area and bottle or pot traps that are usually, but not always baited to attract prey, and are periodically lifted out of the water. The Mediterranean Sea, with an area of about of 2. 5 million km (970,000 sq mi), is shaped according to the principle of a bottle trap. It is easy for fish from the Atlantic ocean to swim into the Mediterranean through the narrow neck at Gibraltar, and difficult for them to find their way out. It has been described as "the largest fish trap in the world".

The prehistoric Yaghan people who inhabited the Tierra Del Fuego area constructed stonework in shallow inlets that would effectively confine fish at low tide levels. Some of this extant stonework survives at Bahia Wulaia at the Bahia Wulaia Dome Middens archaeologicalsite. In southern Italy, during the 17th century, a new fishing technique began to be used. The trabucco is an old fishing machine typical of the coast of Gargano protected as historical monuments by the homonym National Park. This giant trap, built in structural wood, is spread along the coast of southern Adriaticespecially in the province of Foggia, in some areas of the Abruzzese coastlines and also in some parts of the coast of southern Tyrrhenian Sea.

Indigenous Australians were, prior to European colonisation, most populous in Australia's better-watered areas such as the Murray-Darling river system of the south-east. Here, where water levels fluctuate seasonally, indigenous people constructed ingenious stone fish traps. Most have been completely or partially destroyed. The largest and best-known are those on the Barwon River at Brewarrina, New South Wales, which are at least partly preserved. The Brewarrina fish traps caught huge numbers of migratory native fish as the Barwon River rose in flood and then fell.

In southern Victoria, indigenous people created an elaborate system of canals, some more than 2 km long. The purpose of these canals was to attract and catch eels, a fish of short coastal rivers (as opposed to rivers of the Murray-Darling system). The eels were caught by a variety of traps including stone walls constructed across canals with a net placed across an opening in the wall. Traps at different levels in the marsh came into operation as the water level rose and fell. Somewhat similar stone-wall traps were constructed by native American Pit River people in north-eastern California.

A technique called dam fishing is used by the Baka pygmies. This involves the construction of a temporary dam resulting in a drop in the water levels downstream— allowing fish to be easily collected. Also used in Chile, mainly in Chiloé, which were unusually abundant (fish weir and basket fish trap).

TYPES AND METHODS

The manner in which fish traps are used depends on local conditions and the behaviour of the local fish. For example, a fish trap might be placed in shallow water near rocks where pikes like to lie. If placed correctly, traps can be very effective. It is usually not necessary to check the trap daily, since the fish remain alive inside the trap, relatively unhurt.

FISHING TACKLE

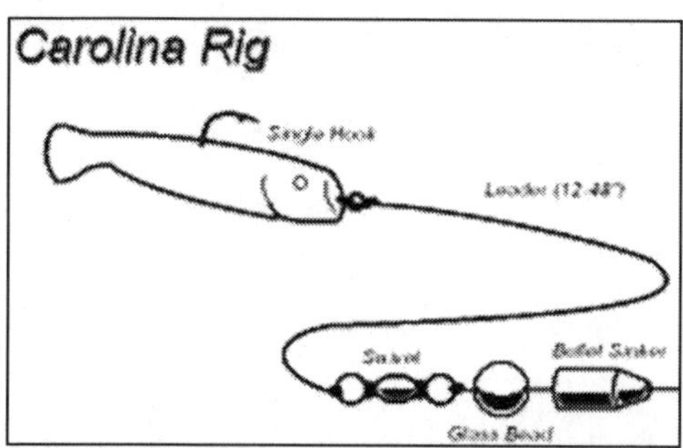

Fig. A completed assembly of tackle ready for fishing is sometimes called a rig, such as this Carolina rig.

Fishing tackle is the equipment used by fishermen when fishing. Almost any equipment or gear used for fishing can be called fishing tackle. Some examples are hooks, lines, sinkers, floats, rods, reels, baits, lures, spears, nets, gaffs, traps,waders and tackle boxes. Gear that is attached to the end of a fishing line is called terminal tackle. This includes hooks, leaders, swivels, sinkers,floats, split rings and wire, snaps, beads, spoons, blades, spinners and clevises to attach spinner blades to fishing lures. Sometimes the term fishing rig is used for a completed assembly of tackle ready for fishing. Fishing tackle can be contrasted with fishing techniques. Fishing tackle refers to the physical equipment that is used when fishing, whereas fishing techniques refers to the manner in which the tackle is used when fishing. The term tackle, with the meaning "apparatus for fishing", has been in use from 1398 AD. Fishing tackle is also calledfishing gear. However the term fishing gear is more usually used in the context of commercial fishing, whereas fishing tackle is more often used in the context of recreational fishing.

SPEARS

Fig. The Filipino Negritos traditionally used bows and arrows to shoot fish in clear water.

Spearfishing is an ancient method of fishing conducted with an ordinary spear or a specialised variant such as a harpoon, trident, arrowor eel spear. Harpoons are spears which have a barb at the end. Their use was widespread in palaeolithic times. Cosquer cave in Southern France contains cave art over 16,000 years old, including drawings of seals which appear to have been harpooned. Tridents are spears which have three prongs at the business end. They are also called leisters or gigs. They feature widely in early mythology and history. Modern spears can be used with a speargun. Some spearguns use slings (or rubber loops) to propel the spear. Polespears have a sling attached to the spear, Hawaiian slings have a sling separate from the spear, in the manner of an underwater bow and arrow. A bow or crossbow can be used with arrows in bowfishing.

TRAPS

Fig. Vietnamese traditional fish trap.

Fishing traps are culturally almost universal and seem to have been independently invented many times. There are essentially two types of trap, a permanent or semi-permanent structure placed in a river or tidal area and pot-traps that are baited to attract prey and periodically lifted. They might have the form of a fishing weir or a lobster trap. A typical trap can have a frame of thick steel wire in the shape of a heart, with chicken wire stretched around it. The mesh wraps around the frame and then tapers into the inside of the trap. When a fish swims inside through this opening, it cannot get out, as the chicken wire opening bends back into its original narrowness. In earlier times, traps were constructed of wood and fibre.

3

Fishing Vessel Techniques

A fishing vessel is a boat or ship used to catch fish in the sea, or on a lake or river. Many different kinds of vessels are used incommercial, artisanal and recreational fishing. According to the FAO, there are currently (2004) four million commercial fishing vessels. About 1. 3 million of these are decked vessels with enclosed areas. Nearly all of these decked vessels are mechanised, and 40,000 of them are over 100 tons. At the other extreme, two-thirds (1. 8 million) of the undecked boats are traditional craft of various types, powered only by sail and oars. These boats are used by artisan fishers.

It is difficult to estimate the number of recreational fishing boats. They range in size from small dinghies to large charter cruisers, and unlike commercial fishing vessels, are often not dedicated just to fishing. Prior to the 1950s there was little standardisation of fishing boats. Designs could vary between ports and boatyards. Traditionally boats were built of wood, but wood is not often used now because it has higher maintenance costs and lower durability. Fibreglass is used increasingly in smaller fishing vessels up to 25 metres (100 tons), while steel is usually used on vessels above 25 metres.

FISHING VESSELS

A fishing vessel is a boat or ship used to catch fish in the sea, or on a lake or river. Many different kinds of vessels are used in commercial, artisanal and recreational fishing. According to the FAO, in 2004 there were four million commercial fishing vessels. About 1. 3 million of these are decked vessels with enclosed areas. Nearly all of these decked vessels are mechanised, and 40,000 of them are over 100 tons. At the other extreme, two-thirds (1. 8 million) of the undecked boats are traditional craft of various types, powered only by sail and oars. These boats are used by artisan fishers.

It is difficult to estimate how many recreational fishing boats there are, although the number is high. The term is fluid, since most recreational boats are also used for fishing from time to time. Unlike most commercial fishing vessels, recreational fishing boats are often not dedicated just to fishing. Just about anything that will stay afloat can be called a recreational fishing boat, so long as a fisher periodically climbs aboard with the intent to catch a fish. Fish are caught for recreational purposes from boats which range from dugout canoes,kayaks, rafts, pontoon boats and small dinghies to runabouts, cabin cruisers and cruising yachts to large, hi-tech and luxurious big game rigs. Larger boats, purpose-built with recreational fishing in mind, usually have large, open cockpits at the stern, designed for convenient fishing.

The vast majority of commercial vessels load and discharge cargo in the safety of ports; their main function at sea is transportation. A fishing vessel differs in that it is used to hunt, locate, catch, load (and sometimes discharge), as well as process and conserve cargo at sea, all in variable weather conditions. In effect, it is a place of work and is a very specialized vessel which is intended to perform all these well defined tasks. The size, deck layout, carrying capacity, accommodation, machinery and equipment of fishing vessels are all related to its function in carrying out its planned operations.

Factors which influence the design of a fishing vessel may be grouped under the following headings:

- The species, location, abundance and dispersion of the fish resources
- Fishing gear and methods
- Geographical and climatic characteristics of the fishing area
- Seaworthiness of the vessel and safety of the crew
- Handling, processing and stowage of catch
- Availability of finance
- Availability of boatbuilding and fishing skills
- Laws and regulations applicable to fishing vessel design, construction and equipment
- Choice and availability of construction materials
- Economic viability

Because of the inherent variations in each of these factors, the diversity of fishing vessels designs operating around the world is enormous, ranging from 2 metre dug out canoes to factory trawlers exceeding 130 metres in length, with trip durations ranging from a few hours to over a year.

TECHNOLOGICAL DEVELOPMENTS

The size and autonomy of a fishing vessel is largely determined by its ability to handle, process and store fish in good condition on board, and thus these two characteristics have been greatly influenced by the introduction and utilization of ice and refrigeration machinery. Other technological developments - especially hydraulic hauling machinery, fishfinding electronics and synthetic twines - have also had a major impact on the efficiency and profitability of fishing vessels.

In developing countries, fishing operations have been greatly influenced by the introduction and widespread use of the outboard engine.

All these technological developments have not only heavily influenced the design of fishing vessels, but, particularly between 1950-1980, resulted in increased productivity, profitability and competition to the extent that many stocks became fully or overexploited. This situation lead to fisheries management measures which included control of fishing effort, sometimes imposed through length limits on fishing vessels. The effect on fishing vessels was seen with designers attempting to maximize the vessel's fishing capacity while maintaining its length within limits. Factory trawler

Fig. FAO/FIIT/J. Turner.

WORLD FISHING FLEET

In 2002 the world fishing fleet numbered about four million vessels: about one-third were decked while the remaining two-thirds were undecked (generally less than 10 m in length). Of the latter, 65 percent were not fitted with mechanical propulsion systems. There is little information available for the undecked/non-motorized vessels but it is estimated that Asia accounts for over 80 percent of them.

The average size of decked vessels remains about 20 GT (around 10-15 metres). Those larger than 100 GT (or longer than 24 m) amounted to about 1 percent of the world fishing fleet. China has approximately 50 percent (25 600) of these larger vessels, while no other country has more than 10 percent of this fleet and about 10 countries together account for 80 percent of the total.

The first-hand statistics on fisheries employment are scarce, incomplete and of low quality. According to FAO records, employment in the primary capture fisheries and aquaculture production sectors in 1998 was estimated to have been about 37 million people, including full-time, part-time and occasional workers. About 60% of them are employed in marine fisheries. About two-thirds of these fishermen work onboard fishing vessels of less than 12 m in length, both decked and undecked.

Safety of the vessel and its crew are considered a paramount design consideration. Despite this fact, there is no international instrument in force concerning the safety of fishing vessels. International conventions and agreements awaiting ratification which concern safety at sea are almost exclusively aimed at vessels 24 metres in length and over (which, in terms of numbers, constitute only about one percent of all vessels in the world's fishing fleet), and therefore do not apply to artisanal vessels and transport boats in developing countries.

Safety regulations for all fishing vessels are left almost entirely to national discretion.

CODE OF SAFETY

Fortunately, however, in 1999 IMO invited FAO and ILO to cooperate in the revision of Part B of the Code of Safety for Fishermen and Fishing Vessels, for vessels of 24m in length and over, as well as, the Voluntary Guidelines for the Design, Construction and Equipment for Small Fishing Vessels, that is vessels of 12m in length and over but less that 24 m in length. In addition Part A of the Code was included in the revision. The revised Code and the Voluntary Guidelines were approved by IMO in 2004. At the twenty-sixth session of the Committee on Fisheries in March 2005, FAO welcomed the revised Code and Voluntary Guidelines and recommended the early publication by IMO of these documents. The Governing Body of ILO approved the revised texts later the same year.

In December 2004, IMO agreed to include in the work programme of the Sub-Committee on Stability and Load Lines and on Fishing Vessel Safety (SLF) a new high priority item on "Safety of small fishing vessels". The aim being to develop safety recommendations for decked fishing vessels of less than 12m in length and undecked vessels of any length, bearing in mind that the largest majority of fishing fatalities occur aboard such vessels

In September 2005, the SLF Sub-Committee reviewed a document submitted by FAO that outlinesideas relating to the development of the new safety standards and confirms FAO's commitment to the exercise. The Sub-Committee established an intersessional correspondence group and approved a timeframe for the development of the safety standards with a target completion date of 2009.

In 2006, it was agreed that the title of the proposed new standards should be: Safety recommendations for decked fishing vessels of less than 12 metres in length and undecked fishing vessels. The purpose of the safety recommendations is to provide guidelines to competent authorities for the design, construction, equipment and training of the crew of small fishing vessels.

HISTORY

TRADITIONAL FISHING BOATS

Fig. Viking boat showing clinker planking.

Early fishing vessels included rafts, dugout canoes, and boats constructed from a frame covered with hide or tree bark, along the lines of a coracle. The oldest boats found by archaeological excavation are dugout canoes dating back to the Neo-lithic Period around 7,000-9,000 years ago. These canoes were often cut from coniferous tree logs, using simple stone tools. A 7000-year-old seagoing boat made from reeds and tar has been found in Kuwait. These early vessels had limited capability; they could float and move on water, but were not suitable for use any great distance from the shoreline. They were used mainly for fishing and hunting. The development of fishing boats took place in parallel with the development of boats built for trade and war. Early navigators began to use animal skins or woven fabrics for sails. Affixed to a pole set upright in the boat, these sails gave early boats more range, allowing voyages of exploration.

Around 4000 B. C. , Egyptians were building long narrow boats powered by many oarsmen. Over the next 1,000 years, they made a series of remarkable advances in boat design. They developed cotton-made sails to help their boats go faster with less work. Then they built boats large enough to cross the oceans. These boats had sails and oarsmen, and were used for travel and trade. By 3000 BC, the Egyptians knew how to assemble planks of wood into a ship hull. They used woven straps to lash planks together, and reeds or grassstuffed between the planks to seal the seams. An example of their skill is the Khufu ship, a vessel 143 feet (44 m) in length entombed at the foot of the Great Pyramid of Giza around 2,500 BC and found intact in 1954.

At about the same time, the Scandinavians were also building innovative boats. People living near Kongens Lyngby in Denmark, came up with the idea of segregated hull compartments, which allowed the size of boats to gradually be increased. A crew of some two dozen paddled the wooden Hjortspring boatacross the Baltic Sea long before the rise of the Roman Empire. Scandinavians continued to develop better ships, incorporating iron and other metal into the design and developing oars for propulsion.

By 1000 A. D. the Norsemen were pre-eminent on the oceans. They were skilled seamen and boat builders, with clinker-built boat designs that varied according to the type of boat. Trading boats, such as the knarrs, were wide to allow large cargo storage. Raiding boats, such as the longship, were long and narrow and very fast. The vessels they used for fishing were scaled down versions of their cargo boats. The Scandinavian innovations influenced fishing boat design long after the Viking period came to an end. For example, yoles from the Orkney island of Stroma were built in the same way as the Norse boats.

SMALL FISHING VESSELS

Small Fishing Vessels are vessels that are used in commercial fishing and are from 0 to 150 gross tonnage, not more than 24. 4 metres in length overall. Vessels, other than sailing ships, that take on loads at sea from vessels engaged in catching or transporting living resources of the sea including fish and marine vegetation are also commercial fishing vessels.

REDUCE THE RISK

Commercial fishing can be a hazardous occupation. The risk can be reduced when a qualified, well trained crew operates a well designed, equipped and maintained vessel in accordance with established rules and procedures.

REGULATIONS ARE THE MINIMUM LEVEL OF SAFETY

Transport Canada regulations are developed in consultation with the fishing industry to provide a minimum level of safety for all these aspects – crew, vessel and operations. The Small Fishing Vessel Inspection Regulations set out the requirements for building and equipping a small fishing vessel while the Marine Personnel Regulationscontain requirements for crew size, certification and training. Other regulations, including the Collision Regulations and theRegulations for the Prevention of Pollution from Ships and for Dangerous Chemicals also apply.

Under the Canada Shipping Act, 2001, vessel owners are responsible for understanding the regulatory requirements that apply to their operation and making sure that the operation is in compliance at all times. Owners are also responsible for developing procedures for the safe operation of the vessel and for dealing with emergencies as well as making sure that crew members receive safety training. Remember that regulations establish a minimum level of safety. You can increase the level of safety on your vessel with equipment, such as personal flotation devices and EPIRBs. Consider the risks and protect yourself accordingly.

SAFETY TRAINING PAYS OFF

Transportation Safety Board investigations have found that in an emergency, crew safety largely depends on the capability and reliability of survival equipment, as well as the crew's familiarity with the equipment and their skill in using it. Crews who are familiar with their vessel's survival gear are better able to respond to an emergency. The Board found that the regular practice of survival drills was a "major factor in the crew's survival. "

Transportation Safety Board Investigation Report M98N006 (Atlantic Prize). Fishing vessel masters are responsible, under the Marine Personnel Regulations (Section 206), for ensuring that each member of the complement becomes familiar with the following before taking part in a voyage,

i. The shipboard equipment that are specific to the vessel,

ii. The operational instructions that are specific to the vessel, and
iii. Their assigned duties; and can effectively perform their assigned duties when performing duties vital to safety or the prevention or mitigation of pollution.

The master is also responsible for keeping a training record of that includes the following information:

i. The name of each person trained,
ii. The equipment they were trained on,
iii. What the training was about, and
iv. The days on which they were trained.

COMMERCIAL VESSELS

Fig. The German factory ship Kiel NC 105.

The 200-mile fishing limit has changed fishing patterns and, in recent times, fishing boats are becoming more specialised and standardised. In the United States and Canada more use is made of large factory trawlers, while the huge blue water fleets operated by Japan and the Soviet-bloc countries have contracted. In western Europe, fishing vessel design is focused on compact boats with high catching power.

Commercial fishing is a high risk industry, and countries are introducing regulations governing the construction and operation of fishing vessels. The International Maritime Organization, convened in 1959 by the United Nations, is responsible for devising measures aimed at the prevention of accidents, including standards for ship design, construction, equipment, operation and manning.

According to the FAO, in 2004 the world's fishing fleet consisted of 4 million vessels. Of these, 1. 3 million were decked vessels with enclosed areas. The rest were open vessels, of which two-thirds were traditional craft propelled by sails and oars. By contrast, nearly all decked vessels were mechanized. Of the decked vessels, 86 percent are found in Asia, 7. 8 percent in Europe, 3. 8 percent in North and Central America, 1. 3 percent in Africa, 0. 6 percent in South America and 0. 4 percent in Oceania. Most commercial fishing boats are small, usually less than 30 metres (98 ft) but up to 100 metres (330 ft) for a largepurse seiner or factory ship. Commercial fishing vessels can be classified by architecture, the type of fish they catch, the fishing method used, or geographical origin. The following classification follows theFAO, who classify commercial fishing vessels by the gear they use.

TRAWLERS

A trawler is a fishing vessel designed to use trawl nets in order to catch large volumes of fish.

- Outrigger trawlers - use outriggers to tow the trawl. These are commonly used to catch shrimp. One or two otter trawls can be towed from each side. Beam trawlers, employed in the North sea for catching flatfish, are another form of outrigger trawler. Medium sized and high powered vessels, these tow a beam trawl on each side at speeds up to 8 knots.
- Beam trawlers - use sturdy outrigger booms for towing a beam trawl, one warp on each side. Double-rig beam trawlers can tow a separate trawl on each side of the trawler. Beam trawling is used in the flatfish and shrimp fisheries in the North Sea. They are medium sized and high powered vessels, towing gear at speeds up to 8 knots. To avoid the boat capsizing if the trawl snags on the sea floor, winch brakes can be installed, along with safety release systems in the boom stays. The engine power of bottom trawlers is also restricted to 2000 HP (1472 KW) for further safety.
- Otter trawlers - deploy one or more parallel trawls kept apart horizontally using otter boards. These trawls can be towed in midwater or along the bottom.
- Pair trawlers - are trawlers which operate together towing a single trawl. They keep the trawl open horizontally by keeping their distance when towing. Otter boards are not used. Pair trawlers operate both midwater and bottom trawls.
- Side trawlers - have the trawl set over the side with the trawl warps passing through blocks which hang from two gallows, one forward and one aft. Until the late sixties, side trawlers were the most familiar vessel in the North Atlantic deep sea fisheries. They evolved over a longer period than other trawler types, but are now being replaced by stern trawlers.
- Stern trawlers - have trawls which are deployed and retrieved from

the stern. Larger stern trawlers often have a ramp, though pelagic and small stern trawlers are often designed without a ramp. Stern trawlers are designed to operate in most weather conditions. They can work alone when midwater or bottom trawling, or two can work together as pair trawlers.

- Freezer trawlers - The majority of trawlers operating on high sea waters are freezer trawlers. They have facilities for preserving fish by freezing, allowing them to stay at sea for extended periods of time. They are medium to large size trawlers, with the same general arrangement as stern or side trawlers.
- Wet fish trawlers - are trawlers where the fish is kept in the hold in a fresh/wet condition. They must operate in areas not far distant from their landing place, and the fishing time of such vessels is limited.

SEINERS

Fig. A seiner fishing for salmon off the coast of Raspberry Island (Alaska) in July 2009.

Seiners use surrounding and seine nets. This is a large group ranging from open boats as small as 10 metres (33 ft) in length to ocean-going vessels. There are also specialised gears that can target demersal species.

- Purse seiners are very effective at targeting aggregating pelagic species near the surface. The seiner circles the shoal with a deep curtain of netting, possibly using bow thrusters for better manoeuvrability. Then the bottom of the net is pursed (closed) underneath the fish shoal by hauling a wire running from the vessel through rings along the bottom of the net and then back to the vessel.

The most important part of the fishing operation is searching for the fish shoals and assessing their size and direction of movement. Sophisticated electronics, such as echosounders, sonar, and track plotters, may be used are used to search for and track schools; assessing their size and movement and keeping in touch with the school while it is surrounded with the seine net. Crows nests may be built on the masts for further visual support. Large vessels can have observation towers and helicopter landing decks. Helicopters and spotter planes are used for detecting fish schools. The main types of purse seiners are the American seiners, the European seiners and the Drum seiners.

- American seiners have their bridge and accommodation placed forward with the working deck aft. American seiners are most common on both coasts of North America and in other areas of Oceania. The net is stowed at the stern and is set over the stern. The power block is usually attached to a boom from a mast located behind the superstructure. American seiners use Triplerollers. A purse line winch is located amidships near the hauling station, near the side where the rings are taken on board.
- European seiners have their bridge and accommodation located more to the after part of the vessel with the working deck amidships. European seiners are most common in waters fished by European nations. The net is stowed in a net bin at the stern, and is set over the stern from this position. The pursing winch is normally positioned at the forward part of the working deck.
- Drum seiners have the same layout as American seiners except a drum is mounted on the stern and used instead of the power block. They are mainly used in Canada and USA.
- Tuna purse seiners are large purse seiners, normally over 45 metres, equipped to handle large and heavy purse seines for tuna. They have the same general arrangement as the American seiner, with the bridge and accommodation placed forward. A crows nest or tuna tower is positioned at the top of the mast, outfitted with the control and manoeuvre devices. A very heavy boom which carries the power block is fitted at the mast. They often carry a helicopter to search for tuna schools. On the deck are three drum purse seine winches and a power block, with other specific winches to handle the heavy boom and net. They are usually equipped with a skiff.
- Seine netters - the basic types of seine netters are the anchor seiners and Scottish seiner in northern Europe and the Asian seiners in Asia.
- Anchor seiners have the wheelhouse and accommodation aft and the working deck amidships, thus resembling side trawlers. The seine

net is stored and shot from the stern, and they may carry a power block. Anchor seiners have the coiler and winch mounted transversally amidships.

- Scottish seiners are basically configured the same as anchor seiners. The only difference is that, whereas the anchor seiner has the coiler and winch mounted transversally amidships, the Scottish seiner has them mounted transversally in the forward part of the vessel.
- Asian seiners – In Asia, the seine netter usually has the wheelhouse forward and the working deck aft, in the manner of a stern trawler. However, in regions where the fishing effort is a labour-intensive, low-technology approach, they are often undecked and may be powered by outboards motors, or even by sail.

LINE VESSELS

Longliners - use one or more long heavy fishing lines with a series of hundreds or even thousands of baited hooks hanging from the main line by means of branch lines called "snoods". Hand operated longlining can be operated from boats of any size. The number of hooks and lines handled depends on the size of vessel, the number of crew, and the level of mechanisation. Large purpose built longliners can be designed for single species fisheries such as tuna.

On such larger vessels the bridge is usually placed aft, and the gear is hauled from the bow or from the side with mechanical or hydraulic line haulers. The lines are set over the stern. Automatic or semi-automatic systems are used to bait hooks and shoot and haul lines. These systems include rail rollers, line haulers, hook separators, dehookers and hook cleaners, and storage racks or drums.

To avoid incidental catches of seabirds, an outboard setting funnel is used to guide the line from the setting position on the stern down to a depth of one or two metres. Small scale longliners handle the gear by hand. The line is stored into baskets or tubs, perhaps using a hand cranked line drum.

- Bottom longliners -
- Midwater longliners - are usually medium sized vessels which operate worldwide, purpose built to catch large pelagics. The line hauler is usually forward starboard, where the fish are hauled through a gate in the rail. The lines are set from the stern where a baiting table and chute are located. These boats need adequate speed to reach distant fishing grounds, enough endurance for continued fishing, adequate freezing storage, suitable mechanisms for shooting and hauling longlines quickly, and proper storage for fishing gears and accessories.
- Freezer longliners - are outfitted with freezing equipment. The holds are insulated and refrigerated. Freezer longliners are medium to large

with the same general characteristics of other longliners. Most longliners operating on the high seas are freezer longliners.

- Factory longliners - are generally equipped with processing plant, including mechanical gutting and filleting equipment accompanied by freezing facilities, as well as fish oil, fish meal and sometimes canning plants. These vessels have a large buffer capacity. Thus, caught fish can be stored in refrigerated sea water tanks and piks in the catch can also be used. Freezer longliners are large ships, working the high seas with the same general characteristics of other large longliners.
- Wet-fish longliners - keep the caught fish in the hold in the fresh/ wet condition. The fish is stored in boxes and covered with ice, or stored with ice in the fish hold. The fishing time of such vessels is limited, so they operate close to the landing place. Pole and line vessels - are used mainly to catch tuna and skipjack. The fishers stand at the railing or on special platforms and fish with polesand lines. The lines have hooks which are baited, preferably with live bait. Caught tuna are swung on board, by two to three fishermen if the tuna is big, or with an automated swinging mechanism. The tuna usually release themselves from the barbless hook when they hit the deck. Tanks with live bait are placed round the decks, and water spray systems are used to attract the fish. The vessels are 15 to 45 metres o/a. On smaller vessels fishers fish from the main deck right around the boat. With larger vessels, there are two different deck styles: the American style and the Japanese style.
- American style - fishers stand on platforms arranged over the side abaft amidships and around the stern. The vessel moves ahead during fishing operation. "Drawing". FAO.
- Japanese style - fishers stand at the rail in the forepart of the vessel. The vessel drifts during fishing operations. "Drawing". FAO.
- Trollers - catch fish by towing astern one of more trolling lines. A trolling line is a fishing line with natural or artificial baited hooks trailed by a vessel near the surface or at a certain depth. Several lines can be towed at the same time using outriggers to keep the lines apart. The lines can be hauled in manually or by small winches. A length of rubber is often included in each line as a shock absorber. The trolling line is towed at a speed depending on the target species, from 2. 3 knots up to at least 7 knots. Trollers range from small open boats to large refrigerated vessels 30 metres long. In many tropical artisanal fisheries, trolling is done with sailing canoes with outriggers for stability. With properly designed vessels, trolling is an economical and efficient way of catching tuna, mackerel and other

pelagic fish swimming close to the surface. Purpose built trollers are usually equipped with two or four trolling booms raised and lowered by topping lifts, held in position by adjustable stays. Electrically powered or hydraulic reels can be used to haul in the lines.

Fig. Japanese squid jigger in Cook Strait.

Fig. Electric lamps on squid jigger, Lakes Entrance Australia.

- Jiggers - there are two types of jiggers: specialised squid jiggers which work mostly in the southern hemisphere and smaller vessels using jigging techniques in the northern hemisphere mainly for catching cod.

- Squid jiggers - have single or double drum jigger winches lined along the rails around the vessel. Strong lamps, up to 5000 W each, are used to attract the squid. These are arranged 50–60 centimetres apart, either as one row in the centre of the vessel, or two rows, one on each side. As the squid are caught they are transferred by chutes to the processing plant of the vessel. The jigging motion can be produced mechanically by the shape of the drum or electronically by adjustment to the winch motor. Squid jiggers are often used during the day as midwater trawlers and during the night as jiggers.
- Cod jiggers - use single jigger machines and do not use lights to attract the fish. The fish are attracted by the jigging motion and artificial bait.

OTHER VESSELS

- Dredgers - use a dredge for collecting molluscs from the seafloor. There are three types of dredges: (a) The dredge can be dragged along the seabed, scooping the shellfish from the ground. These dredges are towed in a manner similar to beam trawlers, and large dredgers can work three or more dredges on each side. (b) Heavy mechanical dredging units are operated by special gallows from the bow of the vessel. (c) The dredger employs a hydraulic dredge which uses a powerful water pump to operates water jets which flush the molluscs from the bottom. Dredgers don't have a typical deck arrangement, the bridge and accommodation can be aft or forward. Derricks and winches may be installed for lowering and lifting the dredge. Echosounders are used for determining depths.
- Gillnetters - On inland waters and inshore, gillnets can be operated from open boats and canoes. In coastal waters, they are operated by small decked vessels which can have their wheelhouse either aft or forward. In coastal waters, gillnetting is often used as a second fishing method by trawlers or beam trawlers, depending on fishing seasons and targeted species. For offshore fishing, or fishing on the high seas, medium sized vessels using drifting gillnets are called drifters, and the bridge is usually located aft. The nets are set and hauled by hand on small open boats. Larger boats use hydraulic or occasionally mechanical net haulers, or net drums. These vessels can be equipped with an echosounder, although locating fish is more a matter of the fishermen's personal knowledge of the fishing grounds rather than depending on special detection equipment.
- Set netters - also operate gillnets. However, during fishing operations

the vessel is not attached to the nets. The size of the vessels varies from open boats to large specialised drifters operating on the high seas. The wheelhouse is usually located aft, and the front deck is used for handling gear. Normally the nets are set at the stern by steaming ahead. Hauling is done over the side at the forepart of the deck, usually using hydraulic driven net haulers. Wet fish is packed in containers chilled with ice. Larger vessels might freeze the catch.

- Lift netters - are equipped to operate lift nets, which are held from the vessel's side and raised and lowered by means of outriggers. Lift netters range from open boats about 10 metres long to larger vessels with open ocean capability. Decked vessels usually have the bridge amidships. Larger vessels are often equipped with winches and derricks for handling the lifting lines, as well as outriggers and light booms. They can be fitted with powerful lights to attract and aggregate the fish to the surface. Open boats are usually unmechanized or use hand operated winches. Electronic equipment, such as fishfinders, sonar and echo sounders are used extensively on larger boats.
- Trap setters - are used to set pots or traps for catching fish, crabs, lobsters, crayfish and other similar species. Trap setters range in size from open boats operating inshore to larger decked vessels, 20 to 50 metres long, operating out to the edge of the continental shelf. Small decked trap setters have the wheelhouse either forward or aft with the fish hold amidships. They use hydraulic or mechanical pot haulers. Larger vessels have the wheelhouse forward, and are equipped with derricks, davits or cranes for hauling pots aboard. Locating fish is often more a matter of the fishermen's knowledge of the fishing grounds rather than the use of special detection equipment. Decked vessels are usually equipped with an echosounder, and large vessels may also have a Loran or GPS.
- Handliners - are normally undecked vessels used for handlining (fishing with a line and hook). Handliners include canoes and other small or medium sized vessels. Traditional handliners are less than 12 metres o/a, and do not have special gear handling, there is no winch or gurdy. Locating fish is left to the fishermen's personal knowledge of fishing grounds rather than the use of special electronic equipment. Non- traditional handliners can set and haul using electrical or hydraulic powered reels. These mechanised reels are normally fastened to the gunwale or set on stanchions close to or overhanging the gunwale. They operate all over the world, some in shallow waters, some fishing up to 300 metres deep. No typical deck arrangement exists for handliners.
- Multipurpose vessels - are vessels which are designed so they can

deploy more than one type of fishing gear without major modifications to the vessels. The fish detection equipment present on board also changes according to which fishing gear is being used.

- Trawler/Purse seiners - are designed so the deck arrangement and equipment, including a suitable combination winch, can be used for both methods. Rollers, blocks, trawl gallows and purse davits need to be arranged so they control the lead of warps and pursing lines in such a way as to reduce the time needed to convert from one type to the other. Typical fish detection equipment includes a sonar and an echosounder. These vessels are usually designed as trawlers, since the power requirement for trawling is higher.
- Research vessels - a fisheries research vessel (FRV) requires platforms which are capable of towing different types of fishing nets, collecting plankton or water samples from a range of depths, and carrying acoustic fish-finding equipment. Fisheries research vessels are often designed and built along the same lines as a large fishing vessel, but with space given over to laboratories and equipment storage, as opposed to storage of the catch. An example of a fisheries research vessel is FRV Scotia.

RECREATIONAL VESSELS

Sport Fishing Boat

Recreational fishing is done for pleasure or sport, and not for profit or survival. Just about anything that will stay afloat can be called a recreational fishing boat, so long as a fisher periodically climbs aboard with the intent to catch a fish. Usually some form of fishing tackle is brought on board, such as hooks, lines, sinkers or nets. Fish are caught for recreational purposes from boats which range from dugout canoes, kayaks, rafts, pontoon boats and small dinghies to runabouts, cabin cruisers and cruising yachts to large, hi-tech and luxurious big game rigs. Larger boats, purpose-built with recreational fishing in mind, usually have large, open cockpits at the stern, designed for convenient fishing.

Big game fishing started as a sport after the invention of the motorized boat. Charles Frederick Holder, a marine biologist and early conservationist, is credited with founding the sport in 1898. Purpose built game fishing boats appeared shortly after. An example is the Crete, in use at Cataline Island, California, in 1915, and shipped to Hawaii the following year. According to a newspaper report at that time, the Crete had "a deep cockpit, a chair fitted for landing big fish and leather pockets for placing the pole. " It is difficult to estimate how many recreational fishing boats there are, although the number is high. The term is fluid, since most recreational boats are also used for fishing from time to time. Unlike most commercial fishing vessels, recreational fishing boats are often not dedicated just to fishing.

Standard aluminum bass boat, withtrolling motor

- Fishing kayaks have gained popularity in recent years. The kayak has long been a means of accessing fishing grounds.
- Pontoon boats have also become popular in recent years. These boats allow one or two fishermen to get into small rivers or lakes that would have difficulty accommodating larger boats. Typically 8–12 ft in length, these inflatable craft can be assembled quickly and easily. Some feature rigid frames derived from the white water rafting industry.
- Bass boats are small aluminium or fibreglass boats used in lakes and rivers in the U. S. for fishing bass and other panfish. They have swivel chairs for the anglers, storage bins for fishing tackle, and a tank with recirculating water for caught fish. They are usually fitted with an outboard motor and a trolling motor.
- Charter boats are often privately operated, purpose-built fishing boats,

and host fishing trips for paying clients. Their size can range widely depending on the type of trips run and the geographical location.

- Freshwater fishing boats account for approximately one third of all registered boats in the USA. Most other types of boats end up being used for fishing on occasion.
- Saltwater fishing boats vary widely in size and can be specialized for certain species of fish. Flounder boats, for example, have flat bottoms for a shallow draft and are used in protected, shallow waters. Sport fishing boats range from 25 to 80 feet or more, and can be powered by large outboard engines or inboard diesels. Boats used for fishing in cold climates may have space dedicated to a cuddy cabin or enclosed wheelhouse, while boats in warmer climates are more likely to be open.

ARTISAN VESSELS

Artisan fishing is small-scale commercial or subsistence fishing, particularly practices involving coastal or island ethnic groups using traditional fishing techniques and traditional boats. This may also include heritage groups involved in customary fishing practices. According to the FAO, at the end of 2004, the world fishing fleet consisted of about 4 million vessels, of which 2. 7 million were undecked (open) boats. While nearly all decked vessels were mechanized, only one-third of the undecked fishing boats were powered, usually with outboard engines. The remaining 1. 8 million boats were traditional craft of various types, operated by sail and oars. These figures for small fishing vessels are probably under reported. The FAO compiles these figures largely from national registers. These records often omit smaller boats where registration is not required or where fishing licences are granted by provincial or municipal authorities.

Artisan fishing boats are usually small traditional fishing boats, appropriately designed for use on their local inland waters or coasts. Many localities around the world have developed their own traditional types of fishing boats, adapted to use local materials suitable for boat building and to the specific requirements of the fisheries and sea conditions in their area. Artisan boats are often open (undecked). Many have sails, but they do not usually use much, or any mechanised or electronic gear. Large numbers of artisan fishing boats are still in use, particularly in developing countries with long productive marine coastlines. For example, Indonesia has reported about 700,000 fishing boats, 25 percent of which are dugout canoes, and half of which are without motors. The Philippines have reported a similar number of small fishing boats. Many of the boats in this area are double-outrigger craft, consisting of a narrow main hull with two attached outriggers, commonly known as jukung in Indonesia and banca in the Philippines.

4

Understanding the Economics of Fishing

INTRODUCTION

Key to understanding the economics of fishing is the concept of Maximum Economic Yield (MEY) which provides a benchmark to compare current with potential economic performance in fisheries. MEY coincides with the level of harvest or effort that maximizes the sustainable net returns from fishing. A MEY harvest is desirable because it is the catch level that enables society to do the best it can with what nature has provided.

By generating the highest possible economic surplus, commonly called the resource rent, the funds can be spent on those goods and services that contribute to overall welfare. For example, in Iceland many of the fisheries generate a resource rent some of which is taxed and used to pay for schools and hospitals. If there were no economic surplus, because of poor fisheries management, the wellbeing of all Icelanders would be diminished. By contrast, the groundfish fisheries of Newfoundland and Labrador off the east coast of Canada have been badly managed. Instead of generating a surplus that could be used to improve the wellbeing of New Foundlanders, the other provinces of Canada have provided billions of dollars in transfer payments in income support to fishers and their families.

Attempts to extend resource use and particularly employment well beyond MEY are common, and often disastrous. Experience in Canada's Atlantic fisheries provides a striking example. Subsidies provided by the Canadian government – with a specific mandate to maximize employment levels in the industry – greatly extended the amount of fishing effort.

Indeed, three decades ago it was '...estimated that Canada's commercial catch in 1970 could be harvested by 40 per cent of the boats, half as much gear and half the number of fishers'. This is wasteful in itself, but dwindling stocks and the eventual collapse of the Atlantic fisheries – in large part due to over-fishing – further increased the government's burden to maintain incomes. In 1990, for example, and before the collapse of the groundfish stocks, self-employed fishers received $1. 60 in unemployment insurance benefits for every dollar earned in the fishery. The subsequent collapse resulted in 'adjustment programmes' that have cost Canadian tax payers billions of dollars. The management structure, stock level and nature and extent of fishing effort that generates MEY depends on a combination of biological and economic factors.

For the resource rent to be maximized, it must also be the case that the fishing fleet uses its fishing inputs in combinations that minimize the costs of harvest at the MEY catch level. In other words, at MEY fishers are at maximum economic efficiency and there is no overcapitalization of vessels or gear. To understand what the MEY target is, we provide a brief review of a population model that is sometimes used in fisheries – the surplus production model. The implications of the model are illustrated in Figure. It shows the yield or net additions to the stock of fish on the vertical axis and the stock of fish on the horizontal axis, also measured in the same units. We also assume, for the moment, there is no uncertainty about the state of nature.

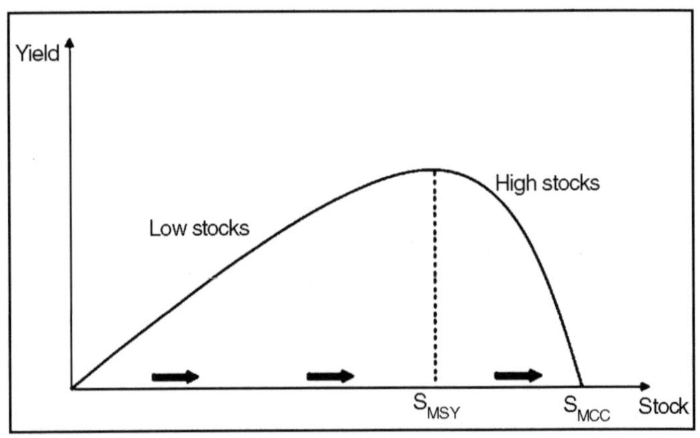

Fig. Surphus-production Model.

The curved line in Figure shows the growth in the stock of fish, or yield, for every possible stock size. It illustrates what is normally referred to as density dependent growth. At low stocks, recruitment is small because there are relatively few fish available to reproduce. Recruitment rises as the stock of fish increases, and then falls as the stock of fish begins to 'crowd' the environment and reaches a maximum limit at its carrying capacity. The stock at maximum carrying capacity thus defines the maximum number of fish (or weight of fish depending on the units we are using) that the environment will support.

With no fishing, the stock of fish will naturally increase – represented by the arrows moving in the right-hand direction to this point. A sustainable harvest occurs when harvest matches yield, or catch is just sufficient to capture new additions to the stock of fish, at any given stock level. In this sense, each point on the yield curve in Figure represents a potential sustainable harvest; with the stock at maximum sustainable yield generating the largest potential catch.

To translate Figure into economic terms, we assume that the price of fish is given – as would be the case for a fishing industry that is competitive and faces a world price for its catch – and, for convenience, we set it equal to one dollar. In this case the yield curve, representing sustainable harvest levels, is also an exact measure of the total revenue from each sustainable catch. We also find it convenient to measure fishing effort (such as nominal days fished or trawl hours) on the horizontal axis, rather than stock size. We note that increases in effort result in a fall in stock such that the two variables generally move in *opposite* directions. Accordingly, Figure measures Total Revenue (TR), which is a function of fishing effort, in dollars on the vertical axis and fishing effort on the horizontal axis.

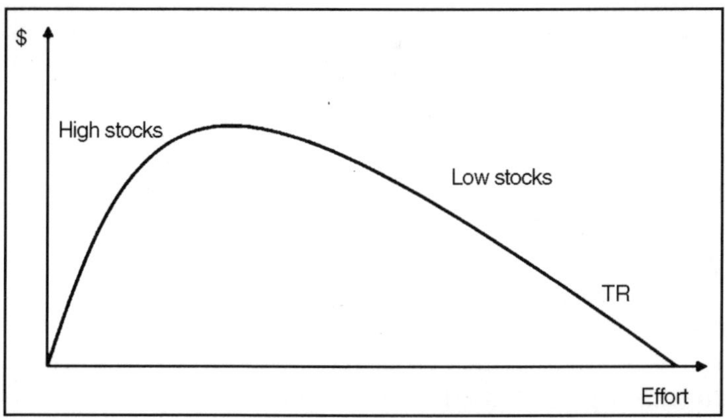

Fig. The Relationship between Total Revenue and Effort

The stock-yield diagram from Figure has thus been flipped 180 degrees to generate Figure. The origin in Figure now represents the fish stock at the maximum carrying capacity because it coincides with zero fishing effort, while the intercept with the largest amount of effort corresponds to a zero stock of fish. By contrast to Figure, a stock of fish that is plentiful, or with 'high stocks', now occurs on the left-hand side of the diagram in Figure and corresponds to low levels of fishing effort, while a low level of fish stocks occurs on the right-hand side of Figure is consistent with high levels of fishing effort.

Nothing yet has been said about the costs of fishing. Again, to keep things simple, assume that all fishing vessels are identical and that the Total Cost (TC) of fishing – including the cost of fuel, crew, bait, gear, etc. and also the opportunity cost of using vessel capital and all other inputs that accounts for the 'normal rate of return' on investment – is proportional to the amount of effort applied.

Under this assumption we can combine TR and TC in one diagram, as in Figure. A key feature of Figure is point B, called the bionomic equilibrium, which corresponds to the level of fishing effort where total fishing cost equals total revenue. This bionomic equilibrium is the point where economic profit (allowing for the opportunity cost of investment and thus is distinct from accounting profit) is zero. If all the economic profit is attributable to the fishery itself, rather than any unique expertise of any individual or group of fishers, this economic surplus is called the resource rent.

Why is point B an equilibrium, or resting point, for the fishery? First, it represents a sustainable harvest. Second, points to the right of effort levels at B will necessarily imply that total costs are larger than total revenues, or that profits are negative.

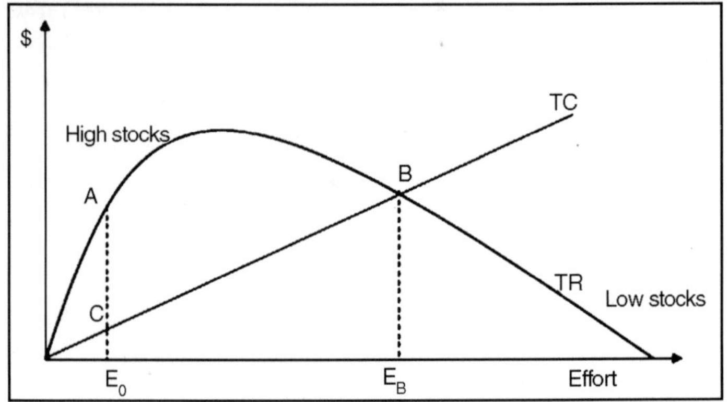

Fig. Bionomic Equilibrium

This must imply that it would be better for firms to employ their capital in their next best alternative. In any case, with negative profits, fishers will eventually go bankrupt and leave the fishery until point B is again obtained. In the case where vessels differ, those that are the least efficient, or have the highest cost of fishing, will be the first to exit. Points to the left of B are more interesting, and illustrate the 'tragedy of the commons' that is associated with every bionomic equilibrium. Begin, for example, at an initial effort level E0, where profits are positive and are measured by the distance A to C. Profits are large in this case because stocks are 'thick' and the cost of fishing is relatively low.

Costs are low at this low level of effort because 'thicker' stocks mean less time is spent fishing and higher catches per unit of effort (each cast of the net, so to speak, catches more fish) lowers the cost per unit of harvest. However, in an unregulated or open access fishery, the existence of positive economic profits – over and above the average rate of return that can be obtained elsewhere – induces new fishing vessels to enter the industry, and those vessels already in the fishery to expand effort and capture the extra profit. As long as profits are positive, this process of effort creep will continue up and until point B, where there is no further incentive to expand effort. When all vessels act in this way, the stock of fish falls over time and the per-unit cost of fishing rises until all economic profits are dissipated.

If any skipper decides to limit fishing effort and conserve stocks, while others do not, that fisher will be relatively worse off. All fishers, acting in their own selfinterest, are induced to fish more, but because those harvesters that increase effort do not take into account the effect of their fishing activity on other vessels in the fishery – including the increased cost of harvest as a result of stock depletion – eventually all vessels are worse off. Indeed, in this sense, point B is undesirable for at least two reasons: first, because profits are zero and the cost of fishing is needlessly high and second, as drawn – and this does not necessarily have to be the case – it would have been possible to obtain the same catch with less effort, lower costs and larger stocks at point A. The case of a bionomic equilibrium makes it clear how profits can be maximized in a fishery, or how to find MEY. In Figure, this occurs at the effort level E* that corresponds to a value of catch $R that creates the largest difference between the total revenue and total cost of fishing. This level of effort maximizes economic profits or the resource rent per unit of effort, and is given by the difference between $R and $C, or R^*. MEY at point A in Figure implies that not only are profits maximized at A, but the value of harvest (both yield in physical terms and the value of catch in terms of revenues) has also increased compared to the bionomic equilibrium at point B.

The reason that profits are now larger at point A is not only because TR is higher, but given that stocks of fish are larger and the amount of 'days' spent fishing is smaller, the cost of fishing is also less, as shown by the move from point B to C. Such a 'winwin' outcome (higher TR and lower TC with a fall in catch) does not always occur with a move from the bionomic equilibrium to MEY.

In many fisheries the cost of fishing may already be sufficiently high (simply rotate the TC curve closer to MEY, implying a fall in effort at MEY) such that a move from a bionomic equilibrium to MEY will require a fall in both the harvest and revenues, but not economic profits. An increase in the price of fish, for example, results in a shift upward of the TR curve at all effort levels, leaving the intercepts unchanged. For a given cost curve, the point of MEY moves closer to MSY, but in this figure never beyond MSY so long as the cost of fishing increases with effort.

In other words, the more valuable landed fish are, the more it pays to work the fishery harder, and thus decrease the equilibrium stock of fish. By contrast, an increase in costs, or a rotation upwards and to the left of the existing TC curve, the new MEY point moves to the left of the previous MEY and lowers optimal fishing effort because with a more costly harvest, it pays to have larger stocks from which to catch. By contrast, technological change that reduces harvesting costs downwards to the right the new MEY point moves to the right of the previous MEY and increases the optimum level of fishing effort. In sum, a fall in the price of fish, or an increase in harvesting costs, implies a lower harvest with less fishing effort and a larger stock size in order to maximize economic profits.

The discussion of MEY underscores the undesirability of MSY (and other biological indicators) as a target, at least as long as having an economically viable fishing industry is an objective. Pursuing MSY alone can result in zero, or even negative, profits at that target level. For instance, it is possible that at a sufficiently high cost of fishing, the fishing effort at MSY can result in the total cost greater than the total revenue. In the long run, no fishing industry will move beyond the bionomic equilibrium given by point B because profits would turn negative (although cases where average costs exceed average revenues for a period of time are not uncommon in poorly regulated fisheries).

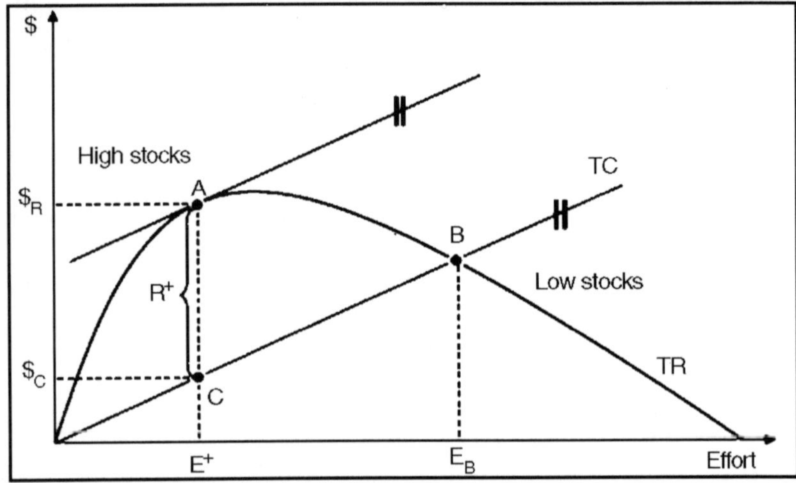

Fig. Maximum Economic Yield (Mey)

However, if a biological target is consistent with negative profits this cannot be a good management goal. Indeed, in the case of a bionomic equilibrium to the left of MSY, where effort is measured on the horizontal axis, a regulatory environment that attempts to target and enforce MSY will simply result in a replication of the bionomic equilibrium. This is clearly undesirable, especially when we consider the considerable resources required to estimate and implement an MSY.

YIELD AND STOCK EFFECTS OF FISHING

Fish stock levels are affected by fishing if the total effort is sufficiently high over some period of time. How much depends on the growth potential of the stock and the total harvest. Change in the stock is expressed by the growth equation,

$$\dot{X} = F(X) - H.$$

From this equation follows,

$$\dot{X} \gtreqless 0 \quad \text{if} \qquad H \lesseqgtr (X).$$

To ensure positive growth of the stock, the harvest must be lower than the natural growth. Biological equilibrium is by definition achieved when X = 0, and in this case equations give,

$$f(E, X) = F(X).$$

Since this is one equation with two variables, X and E, the stock is implicitly given as a function of effort E. This means that at equilibrium the stock level is a function of effort, and from equation it now follows that the equilibrium harvest is also a function of effort. This equilibrium harvest is often called sustainable yield since it can be sustained by the stock for a given level of effort.

We have seen that, knowing the growth function *F(X)* and the short-run harvest function, the sustainable yield may be derived from equation. This can also be done graphically as shown in figure. To simplify the analysis we now assume that the short-run harvest function is linear in effort and stock level:

$$H = qEX.$$

Equation is called the Schaefer harvest function. The parameter q is a constant called the availability parameter. This parameter expresses how effective the effort is in relation to the stock level. If effort is measured in, for example, gill net days, q expresses the ratio between catch per gill net day, H/E, and stock level, X. Thus, the value of q is directly linked to the scaling of E. In some fisheries the combined harvest technology and fish behaviour is such that catch per unit of effort, H/E, is nearly independent of the stock size. In other fisheries catch per unit effort increases with the stock level, but not proportionally as in the Schaefer function.

Panel (a) of figure shows short-run harvest as straight lines for five different effort levels. For the smallest effort E_1 the harvest curve crosses the growth curve for stock level X_1 and harvest H_1. Thus, a small effort – over a sufficiently long time to let the stock reach equilibrium – gives a high stock level and a relatively small catch. A somewhat higher effort level E_2 gives a lower stock level X_2 but a higher sustainable catch, H_2. However, an even higher effort like E_4 gives stock level X_4 that is significantly lower than X_2, even though the sustainable catch H_4 is equal to H_2. Similarly, E_5 gives a catch H_5 equal to E_1, even though the stock level X_5 is much smaller than X_1. The highest possible harvest is reached for effort level E_3 and this harvest is called the maximum sustainable yield (MSY).

The natural-growth stock-level curve in panel (a) has been transformed into a sustainable-harvest effort curve in panel (b). The *H(E)* curve is also called the sustainable yield curve and it connects the long-run harvest potential to fishing effort. This harvest-effort curve has the same form as the growth curve in this case since the Schaefer short-run harvest function is linear in both effort and stock. It is important to note the difference between the short-run harvest function $H = f(E,X)$, depicted as straight lines in panel (a) of figure, and the sustainable yield curve $H(E)$, in panel (b). The former is valid for any combination of effort, E, and stock, X, at any time, whereas the latter is the long-run equilibrium harvest for given levels of effort. The sustainable yield curve is conditional on equilibrium harvest. The main purpose of figure is to derive the equilibrium harvest-effort curve shown in panel (b). Let us now use this to discuss what happens over time if fishing takes place outside equilibrium. Suppose fishers use effort E_1 to harvest a virgin stock at the carrying capacity level K.

To start with, the harvest will be significantly greater than H_1 since the stock level K is bigger than X_1, and this implies that the stock level will decrease. When the stock decreases, the harvest will also decrease until it reaches such a level that, according to the short-run harvest curve designated qE_1X in panel (a) of figure, harvest equals the natural growth of the stock. The decrease in harvest will continue until stock level X_1 has been reached. At this point in time, harvest equals natural growth, and another equilibrium has been established. On the other hand, if fishers use effort E_1 to fish at a stock level lower than X_1 the stock will grow since natural growth is greater than harvest. The length of the transition period between, for example, the virgin stock level K and level X_1 depends on the biological production potential of the stock. Growth curves and sustainable yield curves, as shown in figure may be used to compare different equilibria but cannot be used to tell how long a time the transition from one equilibrium to another will take.

So far in this chapter we have analysed the effects of fishing on a stock with growth compensation. However, if the growth process exhibits depensation or critical depensation, the sustainable yield curve proves to become very different from the case of compensation. This is demonstrated in figures. The former is for the case of depensation and the latter is for the case of critical depensation of growth. In figure panel (a), E_D is the effort that makes the Schaefer harvest curve tangent to the growth curve at the zero stock level. Mathematically, E_D can be found from equation

The left-hand side (lhs) of this equation is the slope of the Schaefer harvest curve, and the right hand side (rhs) is the slope of the growth curve. To ensure a sustainable harvest there is an upper limit on effort which cannot be exceeded, and this effort level is designated E_{MAX} in figures. If effort levels above E_{MAX} are maintained for a sufficiently long time the stock will be biologically over-fished and finally will become extinct. In case of extinction, panel (b) of figures shows that the yield is zero for effort higher than E_{MAX}. Figure panel (b) shows that the harvest curve is double, with an upper and a lower branch for each value of effort between E_D and E_{MAX}. This is due to the existence of two intersection points between each of the linear harvest curves and the growth curve, as shown in panel (a). There is, however, a significant difference between the two branches of the yield curve. The upper part constitutes stable points of harvesting whereas the lower part constitutes unstable harvesting.

An example will explain the stability problem. The harvest curve for effort E_1 intersects with the growth curve for two stock levels, the low one X_{1L} and the high one X_{1H} in panel (a) of figure. For stock levels lower than X_{1L} the harvest curve is above the growth curve and the natural growth is too small to compensate for the harvest. This implies that the stock will decrease from X_{1L} to zero if effort E_1 is maintained over a sufficiently long period of time, indicated in panel (a) by an arrow pointing to the left. Thus, X_{1L} is an unstable equilibrium for the stock harvested by effort E_1.

This would also be the case for all other left-hand side intersections between the harvest curve and the growth curve for effort levels between E_D and $E_{MAX.}$ On the other hand, if the stock level is just above X_{1L} natural growth is larger than harvest for effort E_1 and the stock will increase. An arrow pointing to the right indicates this. Therefore, in this case the stock will in the long run increase towards $X_{1H,}$ which is a stable equilibrium. The lower part of the yield curve in figure panel (b) is dashed to mark that this part represents unstable harvest. Figure shows that, in case of growth with critical depensation, the harvest curve is double for all levels of E between zero and E_{MAX}.

ECONOMIC DEFINITIONS OF PROFIT

In economics, a firm is said to be making a normal profit when total revenues equal total costs. These normal profits then match the rate of return that is the minimum rate required by equity investors to maintain their present level of investment. Economically, the *"normal profit"* is thus treated as a cost, and recognized as one of the two components of the cost of capital.

An economic profit arises when its revenue exceeds the total (opportunity) cost of its inputs, noting that these costs include the cost of equity capital that is met by *"normal profits."* A business is said to be making an accounting profit if its revenues exceed the accounting cost the firm *"pays"* for those inputs. Economics treats the normal profit as a cost, so when deducted from total accounting profit what is left is economic profit (or economic loss).

All enterprises can be stated in financial capital of the owners of the enterprise. The economic profit may include an element in recognition of the risks that an investor takes. It is often uncertain, because of incomplete information, whether an enterprise will succeed or not. This extra risk is included in the minimum rate of return that providers of financial capital require, and so is treated as still a cost within economics. The size of that return is commensurate with the riskiness associated with each type of investment, as per the risk-return spectrum.

"Normal profits" arise in circumstances of perfect competition when economic equilibrium is reached. At equilibrium, average cost = marginal cost at the profit-maximizing position. Since normal profit is economically a cost,

there is no economic profit at equilibrium. In a single-goods case, a positive economic profit happens when the firm's average cost is less than the price of the product or service at the profit-maximizing output. The economic profit is equal to the quantity output multiplied by the difference between the average cost and the price.

Economic profit does not occur in perfect competition in long run equilibrium. Once risk is accounted for, long-lasting economic profit is thus viewed as the result of constant cost-cutting and performance improvement ahead of industry competitors, or an inefficiency caused by monopolies or some form of market failure.

Positive economic profit is sometimes referred to as supernormal profit or as economic rent. The social profit from a firm's activities is the normal profit plus or minus any externalities that occur in its activity. A polluting oil monopoly may report huge profits, but by doing relatively little for the economy and damaging the environment. It would exhibit high economic profit but low social profit.

ACCOUNTING DEFINITIONS OF PROFIT

Note: these definitions are different from those used by economists.

In the accounting sense of the term, net profit (before tax) is the sales of the firm less costs such as wages, rent, fuel, raw materials, interest on loans and depreciation. Costs such as depreciation, amortization, and overhead are ambiguous. Revenue may also be ambiguous when different products are sold as a package, or "*bundled.* " Within US business, the preferred term for profit tends to be the more ambiguous income. Gross profit is profit before Selling, General and Administrative costs (SG&A), like depreciation and interest; it is the Sales less direct Cost of Goods (or services) Sold (COGS),

Net profit after tax is after the deduction of either corporate tax (for a company) or income tax (for an individual). Operating profit is a measure of a company's earning power from ongoing operations, equal to earnings before the deduction of interest payments and income taxes. To accountants, economic profit, or EP, is a single-period metric to determine the value created by a company in one period - usually a year. It is the *net profit after tax* less the *equity charge*, a risk-weighted cost of capital. This is almost identical to the economist's definition of economic profit.

There are commentators who see benefit in making adjustments to economic profit such as eliminating the effect of amortized goodwill or capitalizing expenditure on brand advertising to show its value over multiple accounting periods. The underlying concept was first introduced by Schmalenbach, but the commercial application of the concept of adjusted economic profit was by Stern Stewart and Co. which has trade-marked their adjusted economic profit as EVA or Economic Value Added.

Some economists define further types of profit:

- Abnormal profit (or supernormal profit)
- Subnormal profit
- monopoly profit (super profit)

Optimum Profit: This is the "right amount" of profit a business can achieve. In business, this figure takes account of marketing strategy, market position, and other methods of increasing returns above the competitive rate.

Accounting profits should include economic profits, which are also called economic rents. For instance, a monopoly can have very high economic profits, and those profits might include a rent on some natural resource that firm owns, where that resource cannot be easily duplicated by other firms.

SIGNIFICANCE

Profits are the difference between revenues and costs. In a trade transaction, profit is the difference between the price at which you sell a good and the price at which you bought it. Running a business, net profit is what is left out of turn-over after paying suppliers, workers, financing institution, and the state.

Distributed profits are the income source of the owners of business. As a social group, they are called *"owners"* or *"capitalists"*. The part of the value added not distributed as wages, interests, and taxes, remain within the firm to finance investments.

DETERMINANTS

Maximising profits is said to be the objective of all firms. Indeed, it's not always easy for the management to find out which are the right decisions that would maximise them. For instance, short-run profits can be easily pumped up by avoiding maintenance, discretionary costs, investments, that however are necessary of on-going competitiveness, as you can experiment with this free business game.

Moreover, what maximises the *"overall profits"* is not necessary what allows to attain the maximum of *"profitability"*, *i. e.* the percentage of profits to turn-over, as you can better understand by using this model of monopoly and comparing two policies:

(i) Extremely high prices (= high profitability),

(ii) A price set from a mark-up of 15 per cent on costs.

In reality, firms do have profits targets, and sometimes they pay managers for reaching them, but the goals of firms are broader than profits alone.

Proceeding with other determinants of profits, rising prices of competitors, better sales conditions and skills, a higher overall price level allow for higher prices of the considered firm's products, thus increase nominal profits to the extent that costs are inelastic, *i. e.* they rise less than proportionally to revenues.

Cost structure and its general elasticity to production level is thus relevant to profits. Economies of scale increase profits more than proportionally when sales grow. Conversely, a recession with falling sales levels will hit profits particularly hard in industries where there are economies of scales and high fixed costs.

Rising wages directly reduce profits. If, however, on a macro-economic level, these wages will be spent on domestic goods, higher consumption will boost business revenues, partially counteracting the previous dynamics. Depending on the dynamics of exports and other GDP components, higher wages are compatible with higher profits, as you can see in these real data.

In other terms, productivity gains determine rising profits. High trade profits can prompt other people to entry the market and begin to compete with current traders. In manufacture, this effect, although still present, crucially depends on the easy of imitation of product features and production processes. It's often difficult to enter into highly profitable markets.

If markets were all perfectly competitive in their long run equilibrium, all firms in the economy would have the same constant level of profits: zero. By contrast, in the real world, firms have different profits with certain sectors and certain firms systematically reaching better profits than others. This is due to ubiquitous imperfect competition, barriers to entry, innovation and product differentiation.

PROFITS FROM PROCESS INNOVATION

Consider the case of a competitive market where many firms sell basically the same product at the same price. If they had the same technology and faced the same input prices (*e. g.* wages), they would enjoy similar profit levels. Let's assume that one specialized supplier introduce a new machine that is better than the state-of-art, say a faster machine. The suppler sells it to one of the firms on the market. This will result in lower costs per unit of output, thus higher profits. These profits are used, retrospectively, to pay for the investment in the new machine but after the pay-back period they can be distributed to firm's owners.

The specialized supplier would like to sell again the new machine, either to its first adopter or to its competitors. If not legally restrained, earlier or later, in fact, the competitors will pay attention to the innovation and would like to imitate the first adopter. Slower or faster, innovation will diffuse throughout the market. Possibly some of the reduced costs will reach the consumers in terms of lower prices. If demand is elastic and there are economies of scale a self-sustained positive feedback loop enlarges production capacity, production levels, profits and consumption. If labour is not too weak, also wages will rise. In a certain point on time, firms that did not adopt the new machine will see their profits seriously hit by competition and will have to choose whether to exit the market, to adopt, or to find other competitive advantages. Pressure to reduce wages in failing firms can be an example of these short-term defensive strategy.

PROFITS FROM PRODUCT INNOVATION

Consider a market with product differentiation, as this. An R&D investment over the years leads to an improvement of one product features and the management decides to substitute this new model to the existing one. Let's imagine for simplicity's sake that costs of product are the same as the previous version.

Three main effects will increase sales:

- Consumers who did not buy the good because it did not satisfied their minimum requirements on this feature can now buy, to the extend the improvement is sufficient at their eyes;
- Consumer and positively value the feature could switch from their current provider, to the extend the overall quality of the new good becomes superior;
- Conumer could switch from their current provider, to the extend the price/quality relationships of the new good becomes more convenient.

At the same time, the price of this new version could be set higher than before, so that sales would be broken, unit profits boosted. Overall profits would soar.

IMPACT ON OTHER VARIABLES

Distributed profits go into the income flows of the owner, directly or as dividends. The expected and actual rise in dividends boost the stock exchange quotations. Since shares are concentrated, in their majority, in few hands, and the more so the ownership of private firms, the rich get richer. The demand pattern for consumption goods changes, with a more convex shape. There is an increase in overall consumption quantities, especially of luxury goods, often leading to a more than proportional growth of imports. In certain market segments prices rise. Undistributed profits allows, and high rentability incentives, a new wave of investments, partially funded by banks in this "*euphoria tide*" of positive animal spirits.

Employment in capital goods sectors, as well as in some correlated industries, rise as well. Actual workers can ask for higher wages, credibly manacing strikes that would hurt profits, thus can encounter less-than-usual resistance. Taxation from profits and capital gains soars to give more funds to the State.

LONG-TERM TRENDS

There is no long-term trend in profitability. Absolute profits rise with GDP. The share of profits on value added depends on social groups compromises and conflicts. A certain tendence to a rise of this share can be noticed in Western countries, without being an automatical consequence of development or technology.

BEHAVIOUR DURING THE BUSINESS CYCLE

Profits are extremely pro-cyclical. At the early stages of recovery, inventories go down, with sales surpassing production and injecting new liquidity to the business. The following increase in production is obtained by a higher production utilization of plants and employment. Productivity steeply rise and profits as well.

During recovery and boom, employment and capital accumulation (investments) can go hand in hand, increasing absolute profits if not profitability. At the end of the boom phase, high interest rates on loans (taken for funding investment and discretionary costs) hurt profits, as well as higher wages can do.

Often unexpected, the demand downturn make inventories piling up, freezing resources and forcing a fall in production. Before this adjustment takes place, profits plunge even into negative values. Experiment by yourself the inventory dynamics and how profits react to business cycle through this dynamic monopoly model.

ECONOMIC VALUE OF FISHES

Fishes provide one of the cheapest protein foods. Economic value of fishes can be considered under the major heads: useful fishes and harmful fishes.

USEFUL FISHES

Fishes are useful in many ways; the most important among them are enumerated here.

Fishes as Human Food

Fishes have formed an important item of human diet from the time man appeared on earth. Fish diet provides proteins, fats, Vit, A and D. A large amount of phosphorus and other elements are also present in it. They have a good taste and are easily digestible. Several million tonnes of fish are captured every year by various methods such as spears, baited hooks, traps, nets, etc.

In the order of importance, the principal fresh water fishes consumed as food in India are listed here. The major cultivated fishes allover the world are the fishes belonging to the carp family.

- *Major carps:* Calla calla, Labeo rohita, Labeo calbasu L. fimbriatus, Cirrhina mrigala, C. reba Barbus sp.
- *Cat fishes:* Wallago attu, Mystus seenghala, M. cavasius, BagarulS bagarius, Pangasius sp.

- *Herrings:* Hilsa ilisha, Clupea sF, Setipinna. phasa, Gadusia chapra.
- *Feather backs:* Notopterus chitala, N. notopterus,
- *Live fishes:* Clarias batrachus, Heterpneustes fossilis, chana SF, Anabas testudineus
- *Mullets:* Muglil corsule
- *Miscellaneous:* Labeo bata, L. goniu5', Schizothorax, Tor tor, Berillius, Suntius', Nandus candus, Colisa sp.
- *Exotic species:* Cyprinu5' carpio.

Fish as Food of Cattle

The scrap of canneries as well as entire fishes, that are not relished by man, are dried and ground in mill. This is called fish meal, and is used as artificial food for poultry, pig and cattle. Fish meat is produced in several states like Bombay, Andhra Pradesh, Tamil Nadu, Bengal, Kerala chietly from sardines, mackerels, ribbon fish, etc. The fish is first cooked in large pots containing sufficient quantity of water on fire or on steam. The cooked material is then pressed to remove moisture and dried in the sun on suitable platforms. The resultant product is then stored and if preserved in airtight containers after sterilisation retains it nutritive value for a long time. The fish meal contains about 60 per cent protein and a high percentage of calcium phosphate which is very valuable for cattle and poultry. The manufacture of fish meal can be undertaken as- a cottage industry requiring little expenditure.

Fish Manure

Fishes that are not fit for human consumption are used to prepare fish manure for the fields. During peak season, when there is a large supply of fish, or they are landed in spoiled condition, they are sun dried by spreading them on the beach. The dried fish is ground and converted to manure, which contains a high percentage of nitrogen and phosphate. "Fish guano" is prepared from the material left over after extracting oil from the fish. Mostly sardines are used for preparing fish oil and the waste material forms the "fish guano". It contains 8. 9 per cent nitrogen and phosphate and when mixed with soil forms a rich fertilizer for plants. It is several times richer than ordinary cattle manure

Fish Oil

The most important fishery industry's by-product is fish oil which is of too kirids: liver oil and body oil. The oil extracted from the whole body of the fish is called fish body oil while that obtained tram the liver of certain fish is called the fish liver oil. Liver oil contains Vit A and D, while the body oil contains them in traces only. The refined oil from the liver of fishes has a medicinal value. The body oil has many uses such as in painting, varnishing, soap, candle, leather and steel industries. Liver oil contains 55-75 per cent fat, 5-10 per cent protein and rest is water.

Fish Glue and Isinglass

Liquid glue is prepared from the connective tissue of skin tram head and body of fish. The glue is used as an adhesive for paper, wood, leather and glass. The air bladder of fishes is used for preparing isinglass, which is a shinning powder and is used for clearing wines, bear, making edible jelly and in the preparation of adhesive material. The air bladder is removed from fish, washed in cold water and flattened by beating it on a piece of wood. The bladder is then dried in the sun and is exported for the preparation of isinglass.

Fish Leather

The skin of several fishes like the sharks and rays are used for making polishing and smoothing material. The dried and treated skin is also used for preparing ladies shoes, money bags, suitcases, belts, etc.

Fish Fin

The fins of sharks are exported to China where they are used for preparing soap. Fins are also used for preparing decorative items.

Biological Control

Several species of fishes are larvivorous in habit, and feed upon insect larvae c. g. Chela, Punitivs, Berilius, Danio, Calisa, Raspora, Esomus, Ambassis, Aplocheilus sp. etc. , sev- eral diseases are spread by mosquitoes, hence the larvivorous fish are introduced in the waters of the area. They feed upon larvae and help in reducing the population of mosquitoes.

For Manufacture of Artificial Pearls

The material obtained by scraping the silvery coating of the scales of certain fishes is used for polishing the hollow: glass beads. These beads are then filled with wax and marketed as artificial pearls, used in jewellery.

For Sports and Games

Fishing forms an important outdoor game for million people catch various species of trouts, salmen, carps and other species of fishes.

For Decoration

Several species of beautifully coloured fishes are kept in aquaria, ponds and lakes and used for ornamentation. They are bred and maintained by stockists who offer numerous varieties for this purpose, and earn their livelihood from this trade.

Industries

As the fish forms a rich source of food, millions of people are engaged in fishing industry and depend on fisheries for their livelihood in various ways. Besides those who directly catch the fish for marketing, there are equally large number of people engaged in subsidiary industries like refrigeration, preservation, canning and in manufacture of fish products and by-products.

HARMFUL FISHES

Some species of fish are the intermediate hosts of many parasites causing diseases in man and other animals. Sharks are extremely dangerous in the sea and injure fisherman and damage their nets. Certain species of fishes have poisonous sting or spine (*e. g.* , string ray) and cause painful wounds which might prove to be fatal. Some electric fishes are capable of giving shock to man. Carnivorous fishes eat away the larvae of useful insects. Some species have poisonous flesh and may prove fatal to man.

FISHERIES TODAY -GROWTH, EXPANSION AND TRADE

During the last 25 years, there has been steady growth and expansion in India to claim an important position in the world fish trade.

The growth in population and demand for more fish compelled a more thorough and systematic survey in the early seventies. From 1972-83, the Indian Sea-food Industry, emerged out successfully, ranking as the seventh largest country in the world. According to a recent review, the annual quantity harvested from our seas is 92,000 tonnes, earning about 300 crore rupees in foreign exchange.

According to a survey (1983-84), India has over 20,000 mechanised vessels, 1,40,000 traditional crafts with facilities Ito land in 1,800 centres along its long coast. These facilities include harbour, landing, berthing, and out-fitting facilities, repair houses for boats, proper electricity and water supply, ice plants for refrigeration of fishes, and even roads to connect to the nearest towns and railway stations. The Indian fisheries now employ 4 million employees who include fishermen and their families, technologists, processors, scientists, fishery biologists and exporters.

With the expansion of fisheries markets, increase in production and export, the Government of India faced problems of resources and their quality marketing technology, fish and fisheries education and research, the overall control of sea food industry, etc. The organisation and institutions that the Government has established to regulate and guide the several aspects of fishery industry and allied concerns are varied.

- Central Marine Fisheries Research Institute (Cochin, Kerala state) undertakes research in all aspects of marine fishes and fisheries.
- Inland Fisheries Research Institute (Barrackpore, W. Bengal) deals

with the research on riverine, estuarine, lacustrine (lakes) fishes and their fisheries.

- Indian Institute of Fisheries (Cochin, Kerala state) doing research on ideal crafts and post-harvest technology.
- Institute of Nautical Engineering and Training (Cochin, Kerala state), trains personnels for operating fishing crafts and gears.
- Integrated Fisheries Project of India (Bombay, Maharashtra) concentration on fishes, their biology, exploration of fishing grounds.
- Institute of Fisheries Education (Bombay, Maharashtra) is designed for teaching and research, both theory and practice, on all aspects of fisheries, resources, fishing grounds, crafts and gears, modernisation of boats, trawlers and equipment.
- Marine Products Export Development Authority (Cochin, Kerala state) is the national organisation responsible for the overall developments, regulation and controls of sea food industry as related to export.
- Export Inspection Agency (Cochin, Kerala state) to maintain quality, hygiene, sanitation level of products and inspection of processing plants and package products at export points.
- Fisheries Survey of India (Exploratory fisheries Project) Bombay, (Maharashtra state) to provide training and vital information required for a systematic exploitation of the Indian seas.

Besides the above organisations, study of fishes and fisheries biology has been introduced in colleges and universities in several states of India both in graduate and post graduate courses. There has been a rapid progress in the fisheries industry. Within the last four decades, the industry has reached a top position in exports. In the world sea food trade, India stands first in the prawns and shrimps export tonnage and second largest in marine fisheries, among developing nations, while seventh in the world.

Further, the seas have opened up employment for millions of people. Most of all, thousands of fishermen and their families to have benefited from fishing industry, their earnings have grown and their socio economic conditions have improved.

THE COLLAPSE OF MANY OF THE WORLDS FISHERIES

LEGISLATIVE TRENDS IN CAPTURE FISHERIES

Over-fishing has led to the collapse of many of the worlds fisheries, in many instances because of the existence of open access regimes where the fisher chooses where and when to fish and how much fish to take. Such regimes were justified largely on the erroneous belief that the oceans and other water bodies hold infinite fish resources. In addition, the oceans, in particular the high seas, were *res communis* so that everyone had a right to fish and no one had ownership over the resource or the right to limit access. While many large areas of oceans are now subject to the jurisdiction of coastal states, and high seas fishing is subject to rights and obligations under the 1982 UN Convention, one may be surprised to find that many states have only recently begun to regulate domestic fishing or to consider stringent control over fishing.

In many Southeast Asian states located to the West of the South China Sea, for example, an individual has an unquestionable right to take to the sea to fish, provided that he or she complies with whatever requirements or regulation is in place. In Thailand, an act such as registration of a fishing vessel which is a prerequisite for entering the coastal marine fishery is a mere formality. In this situation, the fisheries management authority in general considers its role to be one in the service or support of fishers or for the promotion and development of fishing.

In Tonga, fishing, particularly small-scale fishing, is considered a livelihood, and no licences have been issued for medium- to large-scale commercial fishing until recently. In many cases government regulation is seen as an unnecessary hindrance to the right to fish. However, the trend today is that governments are realizing that in order to ensure the long-term sustainability of fishery resources, the once open access regime for fishing can no longer continue. To this end, Thailand and Tonga, among others, are considering legislation which could make the right to fish subject to significant controls. Given that it could still be a matter of opinion whether anyone has a right to fish in certain jurisdictions, governments like Iceland enacted fisheries legislation that attempts to clarify the issue as follows:

"Marine resources that are found in Iceland waters and are utilized are the common property of the Icelandic nation. The purpose of this legislation is to ensure the preservation of and sensible utilization of these resources thereby guaranteeing full employment and stable settlement of the country. The issuing of fishing permits, in accordance with this legislation, does not constitute any claims to ownership or irrevocable claims by individual parties over fishing rights".

Senegal makes a similar statement in its 1998 fisheries legislation. Even countries that have never before regulated fisheries, such as Ethiopia, are considering new policies and legislation to properly manage the significant increase in fishery activities and to prevent over-exploitation. As a land-locked country, Ethiopia's fishery activities occur entirely in inland water bodies such as rivers, lakes and reservoirs.

The proposed fisheries laws for Ethiopia set out the basic framework for fisheries management and contain many of the principles referred to above, including the precautionary approach. Other jurisdictions have focused more on asserting the right of the government to manage fisheries resources. Thus Namibia recently stated in legislation that the management, protection and utilization of marine resources in Namibia and Namibian waters shall be subject to its Marine Resources Act of 2001. Cameroon and the Marshall Islands set out the right of the government to manage and control the fisheries resources in stronger terms. A similar approach, in the context of inland fisheries, can be found in Malawi's legislation.

TOWARDS THE USE OF PROPERTY RIGHTS

Related to the trend away from open access to limited access regimes is the move towards creating property rights in fisheries resources and the allocation of such rights. The move from an open access to a limited access regime is in essence a move from one form of administration of property rights to another.

A property rights regime can be a state property regime, private property regime or common (collective) property regime. The most universal is the state property regime, although most states do not commonly depend on "property" or know and administer it as such, but rather on the distribution of fishing rights on the basis of jurisdiction or sovereignty. In such a regime, the state is the custodian of the fishery resource and can decide to leave the resource to free use (open access), can choose to exploit the resource directly through its own agencies or can allot rights to citizens to exploit the resource. The most widely used and practical system of administering the last variant is that the state grants licences to individuals or groups to fish, and in this way controls access to the resources.

At the other end of the scale is the common property regime whereby a local community instead of an individual holds exclusive rights to harvest fish in a certain geographical area.

The developments in Eastern Europe and the former Soviet Union illustrate the effects of the different property regimes. Before the shifts to market economies, fisheries were maintained at an almost constant level by centralized economic plans. However, during the transition period, the fisheries operated under an open access system which stimulated competition but at the same time increased the risk of over-exploitation of the stocks. Therefore, new fishing regulations have been established that set out fish licensing and catch quota regimes, to minimize the risk of collapse. For example, Lithuania enacted a Law on Fisheries in 2000.

Under the private property regime, normally the authorized user who has received a licence has a personal right to fish or harvest and such right is renewed regularly. More recently, licensing schemes have begun to permit the authorized user to sell or lease the right to fish. Such a property regime requires the setting of the total allowable catch (TAC) for a specific fishery, and establishes systems for allocation of the TAC, the transferability and leasability of the rights and the manner in which such rights can be enforced. Iceland has institutionalized the individual quota system in legislation, as has

New Zealand. In Africa, Angola and Mozambique have legislative provisions that will enable the use of a property rights system in the future.

FROM DEVELOPMENT TO SUSTAINABLE UTILIZATION

Fisheries management prior to Rio generally emphasized optimum utilization of fishery resources. In respect of marine capture fisheries, this perspective is reflected in the 1982 UN Convention, which calls on states to optimally utilize fisheries resources. Many fisheries management authorities at that time focused on building capacity or encouraging entry into the fisheries industry to increase fish production for individual profit, revenue generation or domestic consumption. Fisheries management was also an exclusive mandate, often devoid of considerations of the impact of fishing or fisheries activities on associated matters such as the environment. Conversely, other industries did not consider the impact of their activities on fisheries.

There has been a gradual shift in focus to a more "rounded" approach to fisheries management - one that not only ensures exploitation of the resource for economic gain, but also ensures that the resource is maintained at biologically, environmentally and economically sustainable levels. The shift in focus which has been forged in international fora is slowly becoming entrenched in domestic policy and eventually has found its way into legislation. A few examples demonstrate this. Whereas in the past the statement of purpose in fisheries legislation might refer to allocation of a quota levy, the development of fishery or the management of fishery, in recent times one sees an inclusion of a statement of principles and policy relating to sustainable utilization of fishery resources, as in Namibia, Nauru, New Zealand, Papua New Guinea and South Africa. For example, the Marine Resources Act of Namibia has as its purpose to "provide for the conservation of the marine ecosystem and the responsible utilization, conservation, protection and promotion of marine resources on a sustainable basis; for that purpose to provide for the exercise of control over marine resources; and to provide for matters connected therewith". The South Africa Marine Living Resources Act of 1998 has a similar provision.

While preambular provisions of a fisheries act are usually not enforceable, these statements of purpose are given effect in the law's operational provisions, *i. e.* , in the mechanisms it establishes. More significantly, the strong reference in such laws to conservation of the ecosystem and to long-term sustainable utilization is important for it stamps them as very much of the new era of fisheries management.

Similarly, management action was, in the past, based primarily on scientific (biological) information directly affecting the species or fish stock in question. Little or no consideration was given to the effect of fishing on species associated with or dependent on the target species or on the aquatic environment. National legislation now calls for the effects of fishing on non-target or associated species or the aquatic environment to be considered in determining the types of management measures that should apply in a fishery, and to what extent. Legislation that demonstrates this trend can be found in New Zealand and South Africa. Laws of other jurisdictions may not necessarily refer to environmental protection, as in New Zealand, or allow for environmental impact assessments, as in South Africa. Nevertheless, other management actions required under fisheries laws may have the effect of protecting the environment. Sustainable use of the fish stocks and the protection of the environment are now key elements in the recently developed legislation and institutions of Albania, Hungary, Lithuania and Romania.

A related and interesting feature is an increase in references to the precautionary approach or principle. Such references are viewed as an effort by states to give effect to international fisheries instruments or agreements, particularly principle 15 of the Rio Declaration, the UN Fish Stocks Agreement and the Code of Conduct. Even if the precautionary approach is only mentioned in a preambular provision or a broad policy statement and may at best be used only as an aid to interpretation, its significance should not be underestimated, for it "represents a major change in the traditional approach of fisheries management, which until recently has tended to react to management problems only after they reached crisis levels".

IMPROVED ENFORCEMENT

As countries seek to ensure that only sustainable levels of fishing activity are allowed in zones under national jurisdiction, several innovative mechanisms and legislative approaches to fisheries monitoring, control and surveillance have emerged, in particular to curb illegal, unreported and unregulated (IUU) fishing.

Vessel Monitoring Systems

The use of satellite-based vessel monitoring systems (VMS) is a recent development in fisheries monitoring control and surveillance (MCS). VMS ensures that fishing vessels provide reports in real time. VMS can be seen as a direct response to IUU fishing, in particular fishing that is unreported due to problems with radio reporting systems and other conventional means of reporting vessel positions.

At this time, VMS is focused on position reporting, although other VMS information, namely, sighting reports, catch reports, notifications (entry/exit into the exclusive economic zone, port entry, etc.) and analyses, can also be generated by VMS. VMS is currently considered a complementary tool to conventional MCS tools such as sea and air reconnaissance. VMS is in use or in various stages of trial and implementation in many countries and by regional fishery bodies. In the case of one regional fishery body, the Forum Fishing Agency (FFA), the member countries that are mostly developing countries sought to overcome their individual limited resources for MCS by establishing a regional VMS. The issues in the implementation of VMS which are dealt with in legislation are: requiring the installation of VMS components, namely automatic location communicators or vessel tracking units; protecting VMS components; ensuring confidentiality of VMS information; and using VMS information in fisheries enforcement in courts, particularly the admissibility of VMS information such as vessel positions. Recent legislation on VMS in Europe, North America, Southern Africa and the South Pacific has sought to address these issues.

"Long-arm" Approach to Enforcement

Another mechanism that has emerged in fisheries law enforcement is a provision in national fisheries legislation commonly referred to as the "Lacey Clause", from the Lacey Act of the United States. The provision extends the arm of the law, basically by making it unlawful to import fish that has been taken contrary to the laws of another country. A common example of violation of the laws of another state is the taking of fish without a licence where such licence is required by that states' fisheries legislation.

A typical Lacey Act provision states that anyone who lands, imports, exports, transports, sells, receives, acquires or purchases any fish taken, transported or sold contrary to the law of another state is guilty of an offence and liable to a fine. Such a clause was first adopted in the FFA region by Papua New Guinea in 1994, followed by Nauru in 1997 and Solomon Islands in 1998. New Zealand has introduced in a recent amendment to its principal fisheries legislation a Lacey Act-type clause, which prohibits its nationals from fishing in another jurisdiction in contravention of that jurisdiction's laws. An interesting aspect of the New Zealand legislation is that the prerequisite element of bringing fish into the country is not necessary, although such extra-territoriality applies only to New Zealand nationals and vessels.

Alternatives to Criminal Proceedings

Despite innovative efforts such as enactment of Lacey Act clauses, enforcement of fisheries provisions through criminal laws and procedures has a number of drawbacks. There is a high standard of proof, and there may be difficulties in using evidence generated by VMS due to the hearsay rule. In many jurisdictions, extended delays plague the criminal law system. One solution, now applied in the United States, the FFA region and many civil law countries, is the adoption of civil and administrative processes and penalties for dealing with fisheries offences.

This approach presents the advantages of expedited proceedings, lower standards of proof, possibilities for negotiated settlements and hearings which do not necessarily follow strict rules of evidence. Civil penalty schemes for fisheries violations treat certain violations of fisheries laws as civil wrongs penalized by civil penalties, while the right of the offender to decide that he or she be tried under the normal judicial process is preserved. While only a few countries have adopted administrative proceedings to deal with fisheries offences, many states in the Caribbean, Indian Ocean and South Pacific regions have adopted a system of compounding of offences in order to deal swiftly with fisheries violations.

The main element of such a scheme is that instead of resort to traditional court proceedings, the person in whom powers to compound offences is vested (usually the Minister responsible for fisheries or the chief executive officer in the fisheries administration) decides to accept sums of money - usually not more than the maximum of fines allowed - from the offender if it is believed that an offence has been committed. Other requirements in more recent legislation are that offences may be compounded only with the consent of the alleged offender and that the Minister or chief executive officer may be empowered to release any article seized in relation to the offence. The offender retains the right to have the matter against him or her heard in normal judicial fora.

ECONOMIC IMPORTANCE OF MARINE FISHES OF INDIA

Economically important marine fishes of India are as follows:

SARDINE

The oil sardine, Sardinella longiceps are important species constituting about one third of the total marine fish. The clupeids are chiefly represented by the sardines (Sardinella) sardine (Kowala) etc. Sardins represented the important fishery.

THE MACKEREL FISHERY

The Indian mackerel, Rastrelliger Kanagurta is an important commercial fish. Duck Fishery: The Bombey duck fishery is mere restricted to the west coast, but about 10 per cent of the total also comes from the East cost of India.

CAT FISHES

Marine catfishes of commercial importance belong to families, plotosidae and techysuridae. Important species areplotosus canicus, P. augularis, Tachysuras sona, t. maculates. T. dussaurieri and t. jello. Cat fishes predaceous and carnivorous. Feeding mostly as fish and crustaceans.

THE EELS

These are snake like fishes. Three species of eels muraenesox talabonoides, m.cinerus and Anguilla bengalensis are economically important fishes. These are bottom feeders and the food consist of 90 per cent of fish and 10 per cent prawns and crustaceans.

THE RIBBON FISHES

These are thin, ribbon like, silvery in colour with prominent canine teeth and are also called 'hair tails'. Four species of ribbonfishes are mainly found. These are mainly Trichiurus haumela. T. safala, T. intermedious and T. muticus. These are predaceous, carnivorous and sometimes cannibalistic. These are voracious eaters and they feed a fish.

THE SOLE FISHES

The flat fishes or sole fishes have a dorsoventrally compressed body, with an asymmetry of varying degree, having both the eyes on the left or right side of the hand. These are bottom living fishes. These are represented by psettodes, pseudorhombos, solea, paraplagusia and cynoglossus.

CRUSTACEAN FISHES

Several species of crustaceans including prawns, peneids, lobster and crabs are include in the marine catches. The important species are peneus indicus, p. monodon, p. japanicus, metapeneus dobsoni, M. affinis, M. brevicornis, Palaman panulirus, Scylla servata, Metunus palasicus.

MOLLUSCAN FISHERY

A large number of molluscan that can be utilised in various ways occur abundantly in coastal water. There include lamellibranches, Gastropodes and Cephalopodes. Several species of oysters, Calms, Snails and Cuttle fish are used as food by man.

CHANK

Chank is based on the sacred chank Xancus pyrum (Turbinella pyrum) which is widely used as a trumpel in temples. These are also cut into pieces and used as bangles. Flesh of chank also used as food. Chank shell powder is used in some medicines. Since population is increasing, there is demand for more and more food. It needs more and more attention to increase the percentage. Since India has a large coastal line, the prospect of marine fish culture is bright. But it needs political will, modern technology and a commercial depth of mind towards marine fish culture.

FISHERIES PLAY AN IMPORTANT ROLE IN THE ECONOMY OF INDIA

Fishing is one of the oldest occupations of man and learnt fishing much before he could learn something about agriculture. Fishing has assumed much importance in view of the rapidly growing population and depleting land resources. Fish also provides protein rich food and is also a big source of vitamin A, B, and D. There are about 30,000 species of fist in the world out of which about 18000 are found in India. Fish also forms an important part of diet of the people living in the coastal areas of Kerala, West Bengal, Andhra Pradesh, Tamil Nadu, Maharashtra, Karnataka, Goa, and Gujarat. The fish catch in India is of two types:

Marine Fisheries

It includes coastal, off-shore and deep sea fisheries mainly on the continental self upto a depth of 200 meters.

Inland Fisheries

This includes fishes from rivers, lakes, canals, reservoirs, ponds, tanks, etc. India as also the third largest producer and second largest producer of Island fish in the world. The fish production in India has gone high about 2.5 times in a span of nearly two decades. Fish cultures in India have been playing an important role in the economy of the country. As it helps in augmenting food supply, generating employment, raising nutritional level and earning foreign exchange by export.

The fisheries sector also provides employment to over 11 million people engaged fully, partially or in subsidiary activities pertaining to the sector and an equally impressive segment of the population engaged in ancillary activities and accounts for about one per cent of the total agricultural production in India.

However, the fisheries division of the department of Agriculture has been under taking directly or through state governments various production oriented programmes, input supply programmes and infrastructure development programmes besides formulating appropriate policies to increase production and productivity in fisheries sector.

But the main objectives of fisheries development programmes are:

- Enhancing production and productivity of fishermen, fish farmers and fishing industry.
- Increasing nutritional standard of people through fish production.
- Earning of foreign exchange from export of marine products.
- Improving socio-economic conditions of traditional fisherman.
- Employment generation, and
- Conservation of depleted species of fish.

In recognition of the important role of Inland fisheries in overall production of fish, the government has been implementing two important programmes in inland fisheries since fifth/sixth plans. These are Fish Farmers Development Agencies and National Programme for Fish Seed Development. A network of more than 300 Fish Farmers Development Agencies is functioning now. Under the National Programme for Fish seed production many fish seed. Hatcheries have been commissioned in different part of the country.

A programme was also taken up during the Seventh plan for utilisation of organic waste for aquaculture under which many farms are beings developed. A new scheme for development of reservoir fisheries in cooperative sector was initiated during 1989-90 in collaboration with the National Cooperative Development Corporation for development of 27,000 hectare reservoir area in Gujarat, Maharashtra and Karnataka.

GROWTH OF THE FISHERIES SECTOR IN INDIA

Fisheries is an important segment of our economy. It helps in increasing food supply, generate job opportunities, raising nutritional level and earning foreign exchange. The Department of Animal Husbandry and Dairying has been undertaking directly and through the State Governments and the administrations of the Union Territories various production, input supply and infrastructure development programmes and welfare-oriented schemes besides formulating and initiating appropriate policies to increase production and productivity in the fisheries sector.

To focus attention on the contribution made by the fisheries sector to the country's economy and to guide and encourage the fish farmers in their endeavour to achieve higher yield through scientifically proved aquaculture, fish farmers' day is celebrated in all the States and Union Territories on 10th July every year. It was on this day in 1957 that Dr. Hiralal Choudhary, a pioneer in the induced breeding technique in Indian major craps, achieved this feat. As a result of concerted efforts by the Central and State Governments, fish production has continuously been increasing in the country, reaching a record level of 5.96 million tonnes during 2000-01. It is expected to reach the level of 6.05 million tonnes during 2002-2003 compared to the output of 7.5 lakh tonnes in 1950-51. The fisheries sector is a means of livelihood for a large section of the economically backward population in the country. More than 6 million fishermen and fish farmers in India depend on fisheries and aquaculture for their existence.

Standing

India ranks as the 4th largest producer of fish in the world and the second largest producer of freshwater fish after China. The total foreign exchange earnings of the country from the fisheries sector reached a level of ₹ 5815 crore during 2001-2002. Recognising the importance of inland fisheries in the overall production of fish, the Centre has been implementing a scheme known as Development of Freshwater Aquaculture through the Fish Farmers' Development Agencies (FFDAs). These agencies provide a package of technical, financial and extension support to the fish farmers. A network of 429 FFDAs is functioning now covering all potential districts in the country. Since the scheme's inception in 1973-74 till 2000-01, about 6.00 lakh hectares of water area has been brought under fish culture and 6.99 lakh fish farmers trained in improved aquaculture practices through FFDAs. Over 30135-hectare area has been brought under scientific fish culture and 25,729 fish farmers were trained in 2002-03.

The Department of Animal Husbandry and Dairying under the Ministry of Agriculture, Government of India, has been implementing a Central Sector Scheme and a Centrally-Sponsored Scheme since 1964 to provide infrastructure facilities for landing and berthing of traditional fishing craft and mechanised deep sea fishing vessels.

Growth

The Centre has sanctioned six major fishing harbours, 50 minor fishing harbours and 184 fish landing centres. Out of them six major fishing harbours, 33 minor fishing harbours and 130 fish landing centres have been completed. The remaining 17 minor fishing harbours and 54 fish landing centres are at various stages of construction.

In the 2002-03 proposals of the Andhra Pradesh Government for construction of 11 new fish landing centres at different locations along the State's sea coast at a total cost of about ₹.10 crore have been approved. Once developed, these facilities are expected to benefit about 3480 fishing vessels plying along the coast of Andhra Pradesh. Besides, a proposal of the Orissa Government for construction of a fish landing centre at Balugaon at an estimated cost of ₹ 2.35 crore and a proposal of Andaman and Nicobar Administration for construction of a fish landing centre at Dairy Farm, Junglighat at a cost of ₹ 3.68 crore were also approved during 2002-03. The fish landing centre at Balugaon is expected to benefit about 800-1000 country craft, whereas the fish landing centre at Junglighat Dairy Farm, is expected to benefit about 140 non-mechanised fishing vessels. The infrastructure facilities approved under the scheme during the financial year 2002-03 are expected to generate direct employment opportunities to about 25,000 fishermen. A large number of people will also get indirect employment opportunities. A large number of people will also get indirect employment opportunities once the fishery-based industries in and around the infrastructure facilities are developed.

Welfare

Under the welfare programme for the fishing community, the Government extends financial assistance for group accident insurance, saving-cum-relief and development of model fishermen villages encompassing the components construction of houses, community halls and drinking water facilities. During the 9th Plan, Central assistance was extended to various States and Union Territories for construction of about 37,000 houses. Over 10 lakh fishermen were covered under group accident insurance and 2.5 lakh under the saving-cum-relief component annually. In 2002-03, 7148 houses were sanctioned to them. The insurance cover was extended to 10.5 lakh fishers in 2002-03.

INTERNATIONAL TRADE IN AQUACULTURE PRODUCTS

International seafood exports reached US$ 52.9 billion in 1999, a large increase from US$ 35.8 billion in 1990. The share of exports from developing countries has grown from 44 per cent in 1990 to 49 per cent in 1999 and net receipts of foreign exchange rose from US$ 10.4 billion to US$ 16.7 billion in the same period. The rapid growth in aquaculture production has made the sector important to the economy of many developing countries and, in the case of some traded aquatic products; the sector has become either an important source of supply or the main supplier. In these cases, fluctuation in production of farmed products has significant impact on price trends. In general, however, aquaculture products have helped to stabilise supplies of traded products and to bring down prices over the years. This has made what were previously luxury products available at lower prices and helped expand markets.

The extent of regional and international trade in aquaculture products is difficult to analyse because trade in many aquaculture products is not yet well documented in the main producing countries, and since international trade statistics do not distinguish between wild and farmed origin. Thus, the exact breakdown in farmed and wild origin in international trade is open to interpretation. This situation will change gradually, as producers associations emerge in main producing countries and begin to keep records, and in response to various trade regulations/pressures which distinguish between farmed and fished shrimp. The main traded products from aquaculture in 1999 were shrimp and prawns, salmon and molluscs. Other species showing strong growth in trade are tilapia, sea bass and sea bream.

COMMODITIES

Crustaceans

The most prominent product from aquaculture in international trade is marine shrimp and aquaculture has been the major force behind increased shrimp trading during the past decade. Shrimp is already the most traded seafood product internationally, and about 28 per cent of total production now comes from aquaculture (1999: 1.1 million mt). Since the late 1980s, farmed shrimp has tended to act as a stabilising factor for the shrimp industry. Therefore, the major crop failures in Asia and Latin America during the past years have had an impact on overall supply, demand, prices and consumption trends. Considered a luxury product in most markets, shrimp demand is very dependent upon the economic climate in a country, and consumption and trade in an individual country may show large variations from year to year.

The major markets are Japan, the USA and the European Union (EU), and the largest exporters of farmed shrimp are Thailand, Ecuador, Indonesia, India, Mexico, Bangladesh and Vietnam. Demand for shrimp and prawns is expected to increase in coming years. Asian markets such as China, Korea Rep., Thailand, Malaysia, will expand as local economies grow and consumers demand more seafood. This trend is already reducing the availability of shrimp to traditional importers and will eventually put upward pressure on prices if supplies do not expand. Increase in prices will encourage new entries into shrimp farming, if sustainable methods of production are practised that would help avoid production crashes, national curbs on production expansion, or trade embargoes. Trade in crab species has increased with growing aquaculture production (1999: 103 600 mt).

Finfish

In terms of total aquaculture output, finfish production ranks first with 21.5 million mt produced in 1999, or about 67 per cent of the total production from aquaculture. The major part of this is carps (71 per cent of total finfish production in 1999) which are consumed locally in the producing countries (mainly China and India).

International trade in farmed salmon has increased from virtually zero to close to 1 million mt (2001) in less than two decades. The traded species are mainly Atlantic salmon and, to a much lesser extent, coho salmon, which accounted for 88 per cent and 10 per cent of production in 1999 respectively. Growth in trade has followed the growth in salmon production, as the bulk of production is concentrated in a few countries with limited domestic markets -- Norway, Chile and the UK. Norway is the main exporter of Atlantic salmon, and Chile is the main exporter of coho salmon and second largest exporter of Atlantic salmon. The EU is the main market for Norway with some 70 per cent of exports and Japan and the US are the main markets for Chile with some 55 per cent and 30 per cent of exports respectively. Norway has targeted Asia as the future growth market in addition to further penetration of the European markets, and the Norwegian salmon farming industry has over the last few years spent almost US$ 150 million on international promotion and advertising.

Chilean producers foresee strong growth in the USA, Latin American, European and Asian markets, with the exception of Japan. Contrary to the Norwegian industry, Chile produces a large amount of fillets, which are sent fresh to the US market. Chile's other main market is Japan. The global farmed salmon industry is quickly becoming concentrated with a handful of companies present in all producing countries and in all markets and with strong ties to the feed industry. With increased production volumes, costs and prices have been driven down, and at current levels (US$ 2.60-3.40/kg CIF) salmon has become a relatively medium priced product in international seafood markets.

International trade in trout is much less that in salmon, with exports reaching some 122 000 mt in 1999 out of a total production of farmed trout of 475,000 mt. Consumption is concentrated in trout producing countries, but Norway and Chile have been able to farm particular qualities of large-sized heavy pigmented trout for the Japanese market (Japanese 2001 trout imports: 84 000 mt).

Tilapia

Tilapia is another species which has shown a tremendous growth in output (aquaculture production of tilapia and other cyclids amount to some 1 100 000 mt in 1999). International trade is limited but growing, especially between Central America (Costa Rica, Ecuador and Colombia) and the USA, and between Asian producers (Taiwan PC, Indonesia and Thailand) and the USA and Japan. There is also modest trade between Jamaica and the UK. The biggest exporter, Taiwan PC, supplies Japan with high quality tilapia fillets for the sashimi market and ships frozen tilapia to the American market (2001: exports to US 40 000 mt). Taiwan exports about 70 per cent of its domestic tilapia production. Thailand and Indonesia export less than 5 per cent of their production. Vietnam has also recently entered the world tilapia market, and China exported 12 500 mt to the USA in 2001. Zimbabwe now also produces fresh and frozen fillets for the EU market.

Tilapia is now the third most imported aquaculture product by weight in the USA (2001: total US imports of 56 300 mt), after shrimp and salmon. USA imports have been strongly increasing and are forecast to increase further in the future. Long-term tilapia prices are expected to decrease and this should lead to greater exports to the USA as well as to Europe, which presently is undeveloped as a market for tilapia.

In Europe, the sea bream/sea bass industry intends to copy the success of salmon growers. Production reached 120 000 mt in 2001, most of which are exported, mainly to Italy and Spain. The main exporter was Greece, with about 70 per cent of domestic production exported. Italy was originally almost the only export market for Greek production but as a result of market development efforts, Greek exports have now diversified to new markets such as UK, Germany, and France but also Spain for certain sizes. At the opposite end, trade in fingerlings is from Italy, Spain and France to farms in Greece, Malta and Croatia.

As output of sea bass/sea bream has grown, costs have been driven down, and market prices have been reduced to a third during 1990-2002, from US$ 16/kg to around US$ 4-5/kg. The rapid saturation of the market and the parallel rapid decline in prices (60-70 per cent in ten years, compared to 50 per cent in ten years in the case of Atlantic salmon) is attributed to the much smaller traditional market for these species (only Southern Europe) compared to

Atlantic salmon, lack of diversified products and inadequate market development and promotion. However, the substantial drop in prices of these species is opening new markets and expanding existing ones, although acceptable profit margins can only be sustained at the production end through further improvements in productivity and diversification of products. Like for farmed salmon, the sea bass/bream industry is becoming consolidated and several companies are also now quoted on the stock exchange in Greece and Norway.

American catfish is now the fifth most consumed fish in the USA (0.5 kg/ capita edible weight in 2000). Exports are limited as the production is aimed at the domestic market, whereas imports from Vietnam have rapidly gained market share in the American (2001 exports 7 700 mt) and European markets. The reason for the success of catfish is similar to that of tilapia: strong consumer demand for white, easy-to-prepare fillets.

Seaweed

Farmed seaweed production has been growing in the last decade, (7 million Mt in 1999) and is now 88 per cent of total seaweed supplies. Most of output is utilised domestically for food, but there is growing international trade. China, the major producer, has started exporting seaweed as food to Korea Rep. and Japan. Korea Rep. in turn exports some quantities of Porphyry (red seaweed) and under (brown seaweed) to Japan (total Korea Rep. 1999 exports: 14 000 Mt).

Fig. The Rapid Growth in Aquaculture Production has made the Sector Important to the Economy of many Developing Countries

Significant quantities of Eucheuma (red seaweed), are exported by the Philippines, Tanzania and Indonesia to the USA, Denmark and Japan. Total EU imports of seaweed in 2000 amounted to 61 000 mt. Chile is an important extractor, processor and exporter of agar and carrageenen.

Molluscs

International trade in molluscs is relatively limited compared to total output with less than 10 per cent of total output traded. Major importing markets are Japan, USA and France, and major exporters are China and Thailand. The contribution of farmed products to trade is uncertain.

Farmed mollusc production volumes are fairly evenly split between oysters, clams, mussels and scallops, but international mollusc trade is concentrated in scallops and clams (fresh and frozen). Total fresh and frozen scallop imports have grown from 28 000 mt in 1985 to 73 000 mt in 1999, reaching US$ 552 million. Clam imports have grown from 33 000 mt to 159 000 mt in the same period, valued at US$ 265 million. Mussel imports were showing a downward trend after a peak of 175 000 mt in 1992 to reach 137 000 mt in 1993 and 151 000 mt in 1994. However, in subsequent years mussel imports have been showing an upward trend once again until they reached (1999) the record peak of 212 000 mt or US$ 292 million. Oyster imports have been growing steadily from below 10 000 mt in 1985 to 38 000 mt in 1999, reaching US$ 144 million.

Live Seafood

Asia is rapidly increasing its consumption of live seafood as a result of cultural preferences and growing affluence. The live seafood market is largely restricted to the restaurant trade and to consumers with a relatively high disposable income. Major market expansion is anticipated due to demand in China, but expansion is also expected in Malaysia, Singapore and Taiwan PC, as well as in parts of North America and Europe with large Chinese or Asian communities. The potential for aquaculture to supply the market is promising. The sector is already supplying large amounts of shellfish and limited quantities of grouper, crabs and other species. Technological developments in the culture of preferred live food species will increase the contribution of aquaculture to supplies.

Ornamental Aquaculture and Trade

Annual international exports in ornamental fish are around US$ 200 million, or less than 1 per cent of total world fish trade. However, the total value of wholesale ornamental trade is estimated at close to US$ 1 billion, and retail trade about US$ 3 billion. The importance of ornamental fish trade goes far beyond its mere share in international trade. The sector is an important source of income for rural, coastal and insular communities in developing countries, and frequently a welcome provider of employment opportunities and export revenues. Asia represents today more than 50 per cent of the world supply of ornamental fish. New players such as the Czech Republic and Malaysia are now competing with the traditionally dominant suppliers. The main importers are the United States (24 per cent), Japan (14 per cent) and Europe, particularly

Germany (9 per cent), France (8 per cent) and the United Kingdom (8 per cent). In international trade, in value terms, freshwater species represent about 90 per cent, against 10 per cent for marine species. Freshwater species are mostly farmed whereas marine fish come from the wild. However, marine aquaculture is in strong growth as problems related to the environment and lack of sustainable collection practises make aquaculture a more viable long-term alternative

Seed Supplies

There appears to be significant regional and international trade in seed of cultured aquatic organisms, mainly from aquaculture sources, but this is poorly documented at present in most instances. Mention has been made above of regional trade in Mediterranean sea bass and sea bream, but there is also trade in glass eels (*e.g.*, recent large purchases of European eel elvers by China), post-larvae of various cultured shrimps, Indian and Chinese carps, and others. There is also limited trade (in terms of quantity) in brood stock. Documentation of trade in seed will improve gradually as in response to concerns about spread of diseases and the movement of genetic material.

FISH PRODUCTION IN INDIA

The vast potential for marine fisheries along the Indian coastline is well recognised. Against an estimated 4.5 million tonnes (mt) of potential conventional resources, production has been stagnating around 1.4 –1.7mt. The one million tonne mark was reached in 1971 and the Seventh Plan target of 2mt was never achieved. Combined with inland fisheries, the production was around 2.8mt. The ushering in of the new Ocean Regime resulted in the declaration of an Exclusive Economic Zone (EEZ) bringing in two lakh sq.km under India's extended jurisdiction, but this has in no way led to a quantum leap in marine fish production as anticipated. There is misconception that the living resources of the ocean are inexhaustible. Nearly 95 per cent of the conventional resources occur in the continental shelf water, especially in depth of less than 75 metres along the coasts. Fish production in India for the past few years is shown in table.

Table. Production of Fish (Million Tonnes)

Sl.No	Year	Marine	Inland	Total
1	1971	1.162	0.690	1.852
2	1981	1.445	0.999	2.444
3	1982	1.427	0.940	2.367
4	1984	1.779	1.082	2.861
5	1986	1.716	1.204	2.921

As of today, India holds the eighth position in the world in terms of total fish production and is the first among Commonwealth countries, but it is the

136th among 162 countries in terms of per capita consumption of fish. Our average per capita consumption is very disappointing i.e 3.5kg, against 70kg in Japan, 13kg in Myanmar (Burma), 11kg in Sri Lanka and 10 kg in Bangladesh. However, in Kerala the consumption is 4 times the national average.

EXPORT OF MARINE PRODUCTS

Export of marine production increased from ₹ 2.46; crores in 1950-51 to ₹ 234. 84 crores in 1980-81 and to ₹ 597.85 crores in 1988-89. This increase partly reflects the high average unit value realisation of shrimp which was ₹ 82.8 per kg in 1988-89 as against ₹ 33.50 in 1976, though there has been no significant increase in the total export volume of frozen shrimp.

However, this item alone accounts for nearly 80 per cent of the marine exports. The category-wise exports of marine products in 1988-89 is given in table.

Table. Export of Marine Products in 1988-89.

Sl.No.	Item	Quantity in Tonnes	Value in ₹ Crores
1	Frozen shrimp	5,6835	470.33
2	Frozen lobster tail	1663	23.60
3	Frozen cuttle/fish fillets	8,262	23.43
4	Fresh/ frozen fish	11,234	28.44
5	Frozen squids	16,374	38.09
6	Dried fish	3,633	4.43
7	Shark fins/fish maws	315	5.83
8	Others	1,461	3.68

The trend in market preference especially of the biggest buyer, Japan, is for cultured shrimp, accounting for nearly 35-40 per cent of its intake (1988) of about 2.81akh tonnes from countries such as Taiwan, China, Indonesia, the Philippines and Thailand. India's contribution of cultured shrimp export is hardly 3,800 tonnes (1988) as against 30,000 tonnes from China which in 1985 exported only 6,000 tonnes.

This preference for aquaculture products needs serious consideration while formulating the development strategies.

VAST POTENTIAL UNEXPLOITED

India with a coast-line of about 7,517 km into which many large and important perennial rivers discharge their silt laden water and a continental shelf of 259,00 sq. km in the form of a narrow belt, from the shore line to about 200 metre line, the two arms of the Indian Ocean and large number of gulf and bays, extensive back waters, estuaries, lagoons, swamps along with the entire coast-line has large potential fishery resources that can be rendered suitable for sweet water fish farming and that which can be developed for brackish water fish farming is estimated at 1.62 m ha, 0.65 m ha and 2.02 m ha respectively. India has access to a wide variety of tropical marine produce estimated at 4.5

million tonnes from the Exclusive Economic Zone (EEZ). From this annually renewable resource during 1988-89 only about 1.5 million tonnes were landed. The Indian Ocean is the least fished region.

Presently, more than 90 per cent of the fish landings is from the inshore area. The inshore exploitation is largely done by country craft and small mechanised boats using a variety of gears such as trawl nets, prissiness and gill nets. There are about 20,000 small mechanised boats.

To this were added during the last decade about 170 deep-sea-fishing vessels under various programmes. Further the marine fisheries operations were supported by 5 major, fishery harbours (4 commissioned) 29 minor fishery harbours (17 commissioned) and 102 landing centres (82 commissioned) for the convenience of traditional boats.

The West Coast contributed to 69 per cent of total marine-fish-landings, whereas the East Coast yielded 31 per cent. The potential and present yield from the Indian EEZ are given in table.

Table. Potential and Present Yield (In Million Tonnes)

Sl.No.	Item	Indian Exclusive Economic Zone		
		Potential	Presnt	Exploitation
	Pelagic fishes	1.850	0.754	40.7
1	Demersal fishes	1.095	0.493	45.0
2		0.325	0.236	72.6
3		0.180	0.24	13.3
4		0.500	Negligible	0.2
5		0.520	0.057	10.9

The Indian EEZ has rich demersal and pelagic re- sources of Saurashtra, South West Coast, including Wedge Bank, and in South East Coast between 50 and 400 metre depth zones.

In the EEZ around Andaman and Nicobar Islands, a potential yield of 166,000 tonnes of Tuna and Tuna-like-fishes, 40,0000 tonnes of small shaling fishes, and 20,000 tonnes of Demersal stock is estimated. In addition, a stock of 25,000 tonnes of yellow fin and big eye Tuna as well as 50,0000 tonnes of Skipjack have also been estimated. The Tuna resources around Lakshwadeep are estimated at 90,000, tonnes along with abundant Squids and Cuttle fish to be fully exploited.

The Shrimp resources in the coastal water and beyond are estimated at 150-200 thousands tonnes per annum up to a depth zone of 50 metres. In addition the exploitable Shrimp resources estimated at 20,000 tonnes at a depth of 50-450 metre, is a possible future fishery.

Oceanographic surveys have also indicated seizable deep-sea shrimp resources in the upper continental slopes of our EEZ. Exploitable potential and present yield from the Indian Exclusive Economic Zone is given in table.

Table. Exploitable Potential and Present Yield from the EEZ.

Sl.No	Item	Annual Yield in tonnes		
		Potential	Present Yield	Exploitation
1	Tune	740,000	30,000	40,000
2	Cephalopods	180,000	24,000	150,000
3	Deep sea lobster	12,000	2,000	170,000
4	Miscellaneous	1,335,000	301,000	115,000

NEED OF FISHERY CO-OPERATIVES

A co-operative structure would be suitable for this business, especially in a developing country like India, since the conditions prevailing in fisheries sector are almost comparable to those of agriculture sector, *viz.*, poverty and illiteracy of fishermen and perishability of produce.

They would, thus, need guidance, at all stages in their operations. But this is possible only if they are provided with institutional structure adequately modem and responsive to their needs. A co-operative would be their own organisation. Since activities involved in fishing are complex in nature it requires collective action.

Following are the factors responsible for establishing co-operative structure in fisheries.

- Fishing is restricted to on-shore areas up to 20 metres. The off-shore and deep sea areas beyond 20 metres and 80 metres have not been exploited.
- The crafts and implements used are of rudimentary type which cannot stand the rigours and requirements of off shore or deep sea fishing.
- The fishermen are poor, heavily indebted, or work for contractors, who take as much as 50 per cent of the net sales proceeds from them as charges for hire of boats, net and other fishing equipments.
- Fish is one of the most perishable commodity. Coupled with the fact that the climate is subtropical, most of the catch is consumed in areas located near the coast or in the neighbourhood of the landing places.
- Marketing of fish is mostly in the hands of middle men/traders, who dominate the scene to the disadvantage of the fishermen and the consumers. They exploit them because control over marketing is exercised through financing. The cost of finance is very high and disproportionate to the risk involved.

All these factors have adversely affected the fishing industry and the economic condition of the fishermen.

In the absence of institution or organisation, the fishermen have remained financially weak and poor because of their continued exploitation by the money lenders, traders and contractors. Thus, a co-operative organisation at the primary level for the fishermen is absolutely essential.

GROWTH AND DEVELOPMENT OF FISHERMEN'S SOCIETY

Fishery co-operatives have their genesis in the report of the Royal Commission on Agriculture which suggested that fisheries should be put on sound lines.

In 1944, the Fish Subcommittee of Agricultural Policy Committee recommended that like Japan, both direct and indirect assistance should be given to the industry. In 1946, the Co-operative Planning Committee recommended that state aid for the development of fishing industry should be given largely through co-operative societies.

These societies should give financial assistance to their members, and stock and sell fishing craft at a fair price. They should also undertake marketing functions, *i.e.*, functions involving proper arrangement for handling, assembling, reservation, transport and distribution of fish. The societies should be grouped together under the central societies which should have their headquarters in towns. By 1948, there were 1,599 societies with 1.71 lath members with capital of ₹ 50 laths.

Presently the organisational structure of fishery co-operative in India consists of 7950 primary fishermen co- operatives with a membership of 8 lakhs. There are nearly 68 state and central district federations of these societies.

The total business done in 1907 was ₹ 45 crores. They had an uneven development. Maharashtra, Gujarat. and Karnataka federations accounted for 82 per cent of the total turnover.

In recent years the programme is being developed on a "project basis". Its main features are:

- Intensification of fleshing production through introduction of mechanised boats;
- Supply of mechanised boat on credit to groups of members of co-operatives;
- Supply of oils, nets and other requisites to the fishermen.
- Provision of common facilities and services like boat building yards, ice plant for, cold storages, canning plants, transport vehicle, etc.
- Marketing of fish and fish product; and
- Recovery of loan from sale price of fish and fish products.

These integrated projects are being assisted and financed by NABARD and NCDC. The Fishing Farmers Development Agency set up in 1973-74 is intended to coordinate the function of different agencies connected with the inland fisheries and to popularise and promote intensive and integrated fish culture in tanks and ponds.

The number of FFDA was 184 in 1984. By 1987, NCDC sanctioned seven projects of Andhra Pradesh, Kerala, West Bengal, Karnataka and Gujarat.

SOME ISSUES AFFECTING FUTURE TRADE IN AQUACULTURE PRODUCTS

External Factors

Environmental and social concerns continue to influence farmed shrimp exports to North America and Europe. The importance of sustainable aquaculture with no or limited externalities are forcing exporting countries to adopt more sustainable production practices. The introduction of Code of Aquaculture Practises and of eco-labelling schemes will further increase this trend, even though the latter so far have mostly focused on marine fisheries. Another factor of growing importance for aquaculture producers is the increasing attention paid to animal welfare issues, biotechnology and use of GMOs, and labour conditions in general in developing countries.

Distribution Channels

The growth of modern distribution channels such as super and hypermarkets and the expansion of international retail chains have, in many markets, given an important boost to demand and consumption of products from aquaculture. Important buying requirements to guaranteed quantities and prices as well as promotion activities favour in many instances aquaculture products over wild products. Likewise, advances in production technology has allowed cost and price reductions for many farmed species, and future demand from retail chains is expected to grow further, both in developed and developing countries.

Quality

With growing concern about food safety, increasing efforts have been undertaken to improve the quality of aquaculture products. International codex standards cover aquaculture products, and the introduction of mandatory HACCP requirements for exports to the USA and the European Union in 1997 led to significant improvements in production processes and product quality. Some countries have developed comprehensive HACCP plans for selected aquaculture products; for example, the USA now has plans for catfish, crawfish and molluscan shellfish. In other countries, individual aquaculture producers undertake voluntary certification (ISO 9000 standards) for control as well as marketing purposes.

Labelling

New requirements to consumer information and protection and to labelling and traceability are now being implemented in many markets. The EU introduced new legislation on traceability of foodstuffs, including fish and fishery products from 1 January 2002, allowing marketing of fishery products only if labels include certain data. The labels must now specify if the product is farmed or wild, and country of origin or catch area must be given.

WTO and Tariffs

Despite steady reductions in tariffs on fish and aquaculture products in recent years, tariffs as well as import licenses continue to represent barriers to trade in many countries, as do quotas in some countries. This is especially the case in many fast growing economies in Asia, but important markets such as Japan, the European Union as well as the USA all give competitive advantages to domestic producers of many species, especially in the case of processed products. Although average tariffs on imports from developing countries are now estimated at 4.8 per cent, a cut of 27 per cent from previous levels, the averages hide numerous cases of tariff escalation on value added products. In addition to leading to inefficient use of economic resources in both developed and developing countries, this factor continues to hinder development of the fishing industry in developing countries.

The new multilateral trade negotiations as envisioned by the WTO Ministerial Declaration in Doha in November 2001 should lead to more trade liberalisation, lower tariffs and fewer subsidies in fisheries. At the same time, growing membership in the WTO will also lead to lower tariffs and more trade. The recent entry of China into the WTO will also impact international trade and in particular, third country processing of farmed and wild products.

Feed

Production of many carnivorous species depends on fish feed with small pelagic species as raw material. For some inputs such as fish oil, about 80 per cent is now used for aquaculture and this could set a limit for growth in the sector. For this reason, research is now being carried out on alternative inputs in feed such as soybeans for salmon production.

Food Security

As the major part of total output from aquaculture is consumed internally by producing nations, aquaculture is an important source of seafood. Aquaculture has also become a significant source of foreign currency to many developing nations as the products exported usually are the more valuable ones and destined for markets in the developed world. These revenues allow the countries to import other less costly protein, and as such, aquaculture can be considered important to food security even when the output is exported.

FISH-MARKETING PRACTICES AND STRUCTURE OF MARKETS

To make fish available to consumers at the right time and in the right place requires an effective marketing system. Fishermen who catch fish by labouring overnight (from common-property water bodies) do not usually sell fish in retail markets. At the break of day, they take their catches to places where Nikaries/ Beparies, or retailers, meet them and bargain by the lot. At the landing point, the number of intermediaries is low. Only one or two intermediaries may approach a fisherman. Once bargaining has started, other intermediaries remain at a distance and wait for their turn to deal, should the first intermediary fail to obtain the fisherman's lot. If the first intermediary is unsuccessful, another steps in to bargain for the catch. Normally, the first Nikary/Paiker-retailer does not allow this to happen and secures the lot for himself. No open bidding exists in such a case. Therefore, the poor fisherman often falls prey to the Nikari/Bepari/ Paiker-retailer's crude exploitations.

A fisherman, as a seller, cannot negotiate favourable prices for himself mainly because:

- He meets buyers (intermediaries) one at a time and at different times,
- He cannot keep fish for a long time because the product is highly perishable,
- He has no specific place to sit in the market to sell his fish.

Entry into the market is difficult for fishermen for many reasons, mainly because of strong non-cooperation and resistance from the Paikers/retailers. Thus, it is obvious why fishing communities remain poor or are getting poorer over the years, although they trade an important, necessary and every-day commodity. Markets at the primary catch stage are almost completely non-competitive and therefore, exploitation is high.

A pond-farmer, who sells fish by the lot or by species, faces one or two 'nikaries' in his area. Sometimes, Nikaries/Beparies establish their own exclusive trading areas, where other nikaries do not interfere or compete openly.

Therefore, a fish farmer does not encounter a market with many buyers but rather a situation in which he meets more fellow sellers than buyers. This is particularly the case in remote villages. In areas that are well connected by roads and rail, fish farmers contact wholesalers in secondary or higher secondary markets directly and negotiate prices and quantities of fish with the 'Arat's on their own initiatives. Intermediaries, particularly Aratdars, face competition from other wholesalers, which gives the secondary and higher secondary markets an oligopoly-type structure.

Table. Types of Retail Fish Sellers in the Rural Markets

Fish Seller Category	Av. Sellers Per Market	% of all Sellers
Dry fish seller	6.90	21
Preserved fish seller	2.32	7
Live fish seller	5.56	17
Fresh fish seller		
– Small fish	11.96	36
– Large fish	6.20	19

Fifty-five per cent of fish sellers deal in fresh fish in rural primary markets, 17 per cent sell live fish, and sellers of dry fish constitute 7 per cent. Some vendors of small, fresh fish may be the fishermen themselves, who sell directly to the consumers or to the Beparies. If they are, they sit in open places near the fish market and pay exorbitant tolls to extortionists. Thus they face high transaction costs for selling their fish in the market. Competition is higher in retail markets than in any other kind of market, *e.g.*, secondary or higher secondary markets. Prices of fish are determined by the direct interplay of demand for and supply of fish in retail markets.

Markets at all levels have retailing arrangements, that is, a group of retailers that sell fish to consumers. In major cities like Dhaka, Chittagong, Khulna and Rajshahi and in district towns, city corporations or municipalities manage the retail markets. In general, conditions in urban and rural retail markets are not satisfactory regarding stalls, parking, spacing, sanitation, drainage and management. Bargaining, in terms of eye estimation, is still the common practice for pricing fish. Strict grading, sorting and price tagging are ignored in retailing. Quality of the products and a standard for weighing are not enforced at all. In such a situation, cheating and exploitation are unavoidable. Fair pricing according to grade, size, origin and freshness of the fish may not be possible in the absence of standard norms of marketing practices and a lack of enforcement by legal authorities. Thus, market access is limited for the economically weaker section of the consumers.

FISH-MARKETING CHANNELS

Domestic markets and distribution of fish are dominated by a large number of intermediaries. All fish traded internally and for export pass through private channels. Fish distribution usually involves four levels.

Primary Markets

Markets located in villages, district headquarters or at a crossroads are considered primary markets They are usually near areas where fish are caught. Fishermen bring a variety of fishes (dominated by small fish from both open-water capture and from ponds) to the primary markets. Fifty-two per cent of such primary rural markets are held twice a week, 28 per cent three times a week, and 20 per cent are held daily. Of all these markets, 80 per cent are open during morning-hours, particularly for trading milk, vegetables and fish (Market Survey, M-AEP, 1995) and are attended by a relatively small number of sellers and buyers compared to the usual afternoon markets.

Secondary Markets

The Beparies take the fish bought from the Nikaries/fishermen/primary markets/landing points to the nearest Upazila or riverport markets by road, river or rail to sell to wholesalers or through Aratdars. From these secondary markets/assembly points the distribution of fish moves through different channels to urban markets/higher secondary markets by commissioned agents for wholesalers/Aratdars, or by other kinds of Beparies. Bhairab Bazar, Kuliarchar, Narshindi, Munshigonj, Deborghat (Sunamgonj), and Madaripur are landing points and secondary market centres for freshwater fish species.

Higher Secondary Markets

From secondary markets/fish assembly points, Beparies bring fish to the higher secondary markets serving large areas of consumer/terminal markets. The higher secondary market may consist of one or more wholesale markets or centres, where Aratdars deal in fish. These markets are well connected by road, river and rail. Higher secondary markets have trading connections with several secondary markets. Landing ports like Cox's Bazar, Chittagong, Khulna, Bagerhat, Khepupara, Chandpur and Barisal are higher secondary markets and major landing and marketing centres for bulk quantities of marine and brackish-water fish and shrimp species. Markets in district headquarters can be considered as higher secondary markets that are connected with several secondary markets for the supply of fish.

Market Infrastructure and Physical Facilities

The market for fish is crowded at any level. Infrastructure facilities are important for marketing fisheries products domestically and for the physical development of markets. In cities, towns and river ports, city corporations and municipalities often provide infrastructure facilities in the form of pucca roofs, tinshades, pucca platforms, raised selling places and water connections. These physical facilities are emerging even in primary and assembly markets in rural areas through the initiatives of local governments and market management committees. However, washing and cleaning of spaces dedicated to selling fish, and disposal of wastes and residues do not meet hygienically acceptable standards. Physical facilities need to be improved. Fish traders and market managers need to be educated in maintaining sanitary and hygienic places for handling fish. During the summer and rainy season, fish spoil quickly. This results not only in lower prices but also poses health hazards.

Wholesale fish markets are located at secondary, higher secondary and in terminal markets, which may be located at Upazila headquarters, river ports, towns and in cities. The spaces in these markets are generally inadequate for handling highly perishable commodities like fish. However, ice blocks are available in almost all cities and river ports, but not in all Upazila wholesale markets, where waste is still high due to the deteriorating quality of the products sold. The Ministry of Fisheries and Livestock, through one of its farms, the Bangladesh Fisheries Development Corporation (BFDC), has developed domestic fish marketing facilities (physical infrastructures for fish landings and wholesaling) in the secondary and higher secondary markets (channel v). These are harbour and fish landing centres in the coastal regions of Chittagong (fish harbour for deep see fishing trawlers), Cox's Bazar, Barisal, Khepupara, Khulna, Patharghata, Kaptai, Rangamati and Rajshahi.

Assembly centres with proper facilities for berthing, landing and auctioning, with ice plants, cold storage, freezer storage, freezer plants and transport facilities have been developed in landing markets. A modern fish-landing centre was established in Chittagong in 1994 with Japanese assistance for a mechanised fishing fleet of 2000-3000 boats operating from the Chittagong area. In this centre, about 45,000 t of fish can be handled at a time. Fishermen Cooperative Societies also run a major fish-landing centre in Chittagong. The supply of ice much improved during the 1990s. In Bangladesh, the main processing activities for prolonging the shelf life of harvested fish are drying fish and packing fish in ice.

The process of drying also adds a piquant flavour to the fish, which some consumers like. BFDC has established 11 ice plants with a capacity of producing a total of 197 t of ice blocks a day. These plants are in Chittagong-2, Cox's Bazar-1, Rangamati-1, Dhaka (Pagla)-1, Mongla-1, Patharghata-1, Barisal-2, Rajshahi-1 and Khepupara-1. BFDC owns 11 cold storage units with a storage-capacity of 635 t and four freezer plants with a freezing capacity of 61t (blast freezing 43t, plate freezing 12t). At the private levels, many ice plants are operating in and around Dhaka City, Munshigonj, Comilla and Chandpur. Many ice plants have been established in Cox's Bazar, Khulna, Chittagong, Barisal, Khepupara, Sylhet, Patherghata and in other fish landing areas of the coastal belt. As per 1993 figures, Bangladesh has 227 ice plants with a total capacity of producing 4 280 t of ice per day. Most mechanised fishing boats carry ice on board. During times of peak hilsa harvest, fishermen face a shortage of ice. In the face of an expanding private sector, more ice plants and cold storage plants are needed. Since 1992, BFDC has been directing its development activities towards building further landing facilities and ice plants. Some of the developed landing facilities in Khepupara, Patharghata and Barisal remain unutilised and are not claimed by private traders either. Arrangements should be made to hand over these facilities to the private sectors to utilise them better once they are built.

MARKET AND TRADE ROUTES

The center for the Live Food Fish Trade is located in Hong Kong — the markets consumers contribute $400 million to the estimated $1 billion of the trades global value (Seaweb).

Total imports flowing into Hong Kong included 10153 metric tons, of which 30 per cent was re-exported to mainland China (Montaldi). Other major markets include Singapore, mainland China, and Taiwan (Graham). The primary suppliers of wild caught fish are Indonesia (accounting for nearly 50 per cent of Hong Kong's imports), Thailand, Malaysia, Australia, and Vietnam (Graham). However, Taiwan and Malaysia are leading the charge towards farmed live fish specialising in an industry that "harvested annually has probably been in the billions [Metric Tons](Graham)." Farming live fish is gaining popularity as tastes for live fish are burgeoning across South Asia and countries look to become more and more self sustainable, this is appareant in nations with sizable Chinese populations such as Indonesia and Malaysia.

Hong Kong and China are the dominant markets for the live fish, in addition to other cities in the region that have large Chinese populations, including Singapore and Kuala Lumpur(BusinessWorld Nov. 1999). In Southeast Asia, Singapore alone consumes 500 tons of live coral fish a year(ibid).

Exports from Southeast Asia rose to over 5,000 tons in 1995 from 400 tons in 1989(ibid). However, in 1996, exports declined by 22 per cent(ibid). Indonesia, which accounts for over 60 per cent of the harvest, saw exports falling by over 450 tons(ibid). That same year, other Southeast Asian countries have experienced similar drops in stocks of live coral fish for food(ibid). In 1996, the Philippines' exports were halved, while Malaysian exports declined by over 30 per cent(ibid). These decreases in catch have been due to the excessive amount of fish caught for exports and the degradation of the coral reefs from such procedures.

CORRUPTION IN THE TRADE

The live fish trade is a complex issue that involves many different perspectives, all of which must be considered in trying to approach a solution. While, at first, one may point the finger at the fishermen themselves as the criminals, there are many other factors. One is the economic disparity of many of the communities that take part in cyanide, dynamite, or other illegal fishing practices. 40 per cent of the Filipino population and 27 per cent of the Indonesian population is considered to be in poverty. Many whose livelihood once depended on fishing or agriculture are realising it is more lucrative to participate in illegal fishing activities.

Community members who are not a part of the trade are affected by the activities of these illegal fishers. The cyanide fishers profit by taking away from everyone else's trade and food. (Lowe, 7) 'If people were using poison and my take dropped to only a little, I would accept it,' Puah said. 'But I feel heartsick...I catch nothing at all. I have not caught a big fish in a month so there's no point in going fishing this afternoon.'" (Lowe, 7) The local people are often helpless to protect themselves, as government and law enforcement officials have "open pockets" and are also involved in the trade by turning a blind eye to the illegal actions and receiving a take of the profits. "Culpability in cyanide use cannot be understood apart from the larger structures of corruption that permeates resource extraction throughout Indonesia. The Indonesian State bureaucracy extends from Jakarta down to the village level, and radiates out into villages through kinship connections. It is the factor most tightly correlated with illegal trade in natural resources throughout Indonesia." (Lowe, 8)

CAUSE AND EFFECT

Coral reefs found in the South Pacific are regarded as the "rainforest of the sea" harbouring countless fish species large and small. However, recently the live fish trade has threatened the sanctity of these endangered areas. The Global Coral Reef Monitoring Network has issued a recent report that estimates that 25 per cent of the world's reefs are severely damaged and another third are in grave danger.

The live fish trade is part of this alarming ecological trend caused by the popular use of cyanide which is injected into the coral reefs to stun inhabiting fish so they can be easily caught by nets. It is estimated that since the 1960s, more than one million kilograms of cyanide has been squirted into Philippine reefs alone, and since then the practice has spread throughout the South Pacific. (Moore Online) The live fish trade is only growing, in 1994 the Philippines exported 200,000 kg of live fish; by 2004 the Philippines were annually exporting 800,000 kg annually. (Aguiba Online) Although Asian markets are the primary buyers of live reef fish for food, the recently created U.S. Coral Reef Task Force has concluded that the U.S., is the primary purchaser of live reef fish for aquariums as well as eclectic jewelry. (Moore Online) Even though the use of cyanide in the live fish trade is severely detrimental, one must realise that this issue is multidimensional. Small-scale native fishermen of the small South Pacific coastal communities are the backbone of the live fish trade, and are forced to resort to the illegal use of sodium cyanide due to demand and high prices offered by the industry.

FISHING TECHNIQUES

While live food fish trade can be very profitable for those involved, there are many dangerous aspects to it. Through the use of illegal practices such as cyanide fishing, coral reefs and fish communities are put in grave danger. The process of cyanide fishing involves injecting crushed cyanide tablets and squirting this solution from a bottle towards the targeted fish on top of coral heads. Specifically, the cyanide kills coral polyps, symbiotic algae, and other coral reefs organisms that are necessary for maintaining the health of the coral reef. These damages eventually deteriorate the coral reef and lead it into collapse of the entire coral reef ecosystem. The effect on the targeted fish is disorientation and semi-paralysis. After being squirted with cyanide the fish is easy brought to the surface and kept alive in small, on board container.

Fishermen often understand that this practice is harmful and will offer locals a portion of the fish in order to continue fishing. When ingested, small levels of cyanide accumulate in the system causing weakness of fingers and toes, failure of the thyroid gland and blurred vision. Divers without experience may come in direct contact with cyanide, causing death. Estimates show that since the 1960s, over a million kilograms of cyanide have been squirted into the coral reefs of the Philippines alone (Bryant *et al.*). The harm upon the reefs is coming full circle and having a social impact through the limited fish stocks. As fish are depleted from these fishing techniques, the fishermen are having a more difficult time feeding themselves (Cyanide).

Additionally, the use of explosives may be used as a fishing technique in the live food fish trade. While the majority of these fish do not survive the blast of such explosions, the remaining fish that are only stunned are collected for the live food fish trade. The use of cyanide makes a stronger argument in that the more coral reef fish captured alive, the more lucrative the catch is for the fisherman. Live fish, according to the, fetch five times more than a dead fish (Cyanide). This is why banning the live-fish trade would similarly harm the reefs. Fisherman would resort to the "dead fish trade", be forced to deal in a larger quantity of fish, and the process of dangerous fishing techniques continues.

To illustrate the effects of repeated explosions in the proximity of a coral reef system, roughly one half of the coral reefs in Komodo National Park in Indonesia have been destroyed (Moore). The use of cyanide and explosives in fishing proves to be an effective technique in catching fish, but its powers are indiscriminating and as a result the coral reefs are being held hostage to such practices. The future of the reefs are in question just as are the futures of those who subsist from them because "from a long term perspective, the question of ethics of using the ocean...contains a commitment for future generations".

IMPACTS ON HUMANS

In communities like those in the Philippines and Indonesia, people are participating in the live food fish trade because it is a source of income, or at least a source of temporary income. For some communities this is one of the few income-generating opportunities. Along with the environmental, ecological, and economic consequences of this industry, there are serious health risks as well. Because of inadequate training and lack of quality equipment, divers, especially young men are in large risk of paralysis.

SUSTAINABLE PRACTICES

Because of the great profitability of this industry, there is a great incentive to identify sustainable practices. The Marine Aquarium Council (MAC) works to offer the hobbyist with a product that is certified as environmentally sound and sustainable. Additionally, the International Marinelife Alliance (IMA), The Nature Conservancy (TNC), and MAC are working with the Hong Kong Chamber of Seafood Merchants to develop standards for the live fish trade. The Hong Kong Seafood Merchants represent ninety per cent of the buyers of live reef food fish in Hong Kong and have an extensive impact on collection practices.

AQUACULTURE

In an effort to address the damage inflicted on coral reef eco-systems and fish stocks, aquaculture is being utilised to reduce pressure on coral reefs. However, initial efforts to farm grouper have met with significant challenges. There are difficulties with fragile grouper seed that can make it more expensive than wild caught larvae, which can affect natural replenishment rates. Additionally, there are problems with finding suitable food, disease and cannibalism (Johannes and Ogburn). Efforts are also being made in regards to the aquarium fish trade. Juvenile fish are being captured and raised specifically for the industry.

However, there are debates as to whether this practice will affect replenishment rates. "The age of the juveniles is pivotal to the debate, harvesting of postlarvae from the water column is considered to have a much lower (negligible) impact on rates of replenishment than the removal of the larger juveniles from benthic habitats because the postlarvae have yet to undergo severe mortality" (Bell, Doherty and Hair). If studies determine that the capture of juveniles is sustainable, it may help in mitigating the damage from cyanide fishing.

The should also be noted that aquaculture production, specifically grouper rearing is rapidly expanding in Asia. From 1998 to 2001 the Indo-Pacific countries involved in aquaculture; China, Indonesia, Republic of Korea, Kuwait, Malaysia, Philippines, Singapore, and Thailand witnessed a 119 per cent increase in output.

The explosion in this practice can most likely be attributed to the large profit margins that can be derived in very little time. It is estimated that the majority of farms after annual returns can be paid back in less than one year (Siar *et al.* 2002). In comparison to other species of fish such as the Milkfish, the Grouper, because of high demand is able to garner high rates of return, in order to earn 1,000 dollars a grouper farm would only have to raise 400 kilograms in contrast to 5,000 kilograms of Milkfish.

CONCLUSION

India, with its long sea coast stretching 6100 km, numerous rivers, fresh water and brakish water lakes, reservoirs, tanks, ponds and swamps has very high potential for fish and other forms of aquaculture. It has 48 million hectares of sea area, 1. 62 million hectares of fresh water area and 2. 02 million hectares of brakish water area which are available for fishing and fish culture. What perhaps lacking is scientific knowledge about the fish and other forms of aquaculture. We cannot develop these tremendous natural water resources, if we depend only on the old and traditional ways; we need to acquire modern techniques to manage the water resources for fish and other aquaculture.

5

Sustainability Fisheries System

The concept of sustainability has been brought to the Centre of socio-economic and environmental debate after the well-known definition of sustainable development by World Commission on Environment and Development. However, since then the notion of sustainability has generated tremendous interest and an avalanche of publications, even though it has never found an agreed definition of sustainability itself.

Most of sustainability definitions originate from the relationship between humans and natural resources system. Wimberly states that to be sustainable is to provide for food, fibre and other natural and social resources needed for the survival of a group-such as a national or international society, an economic sector, or residential category-and to provide in a manner that maintains the essential resources for present and future generations. Norton argues that sustainability is a relationship between dynamic human economic systems and larger, dynamic, but normally slower changing ecological systems, such that human life can continue indefinitely, human individuals can flourish, and human cultures can develop-but also a relationship in which the effects of human activities remain within bounds so as not to destroy the health and integrity of self-organising systems that provide the environmental context for these activities.

Constanza, in line with the above definitions, defines sustainability in systems properties, stressing that sustainability...implies the system's ability to maintain its structure (organisation) and function (vigour) over time in the face of external stress (resilience). Moreover, more on economic dimension, Solow argues that the system is to be said sustainable as long as the total capital of the system is equal or greater in every next generation. Constanza and Daly mentioned that sustainability only occurs when there is no decline in natural capital.

From several definitions of sustainability, there is one common component in all of them. There is something about maintenance, sustenance, continuity of a certain resource, system, condition, relationship between components of the system, in keeping something in a certain level, and of avoiding decline.

In fishery sciences, the sustainability has long been in discussions and debates on the concept of sustainable yield. Sustainability in fishery has been determined predominantly by the level of sustainable catch which Charles called as the conservation paradigm of sustainable fisheries. Furthermore, it was elaborated that the conservation paradigm in sustainable fisheries makes the definition of sustainability as one of long-term conservation, so that any activity is judged sustainable if it protects the fish stocks without considering much on human-oriented fishery objectives. The icon of MSY (maximum sustainable yield) is one of the very well-known parameter that has been emphasised by the conservation paradigm.

Another perspective, called the rationalisation paradigm, has challenged the dominancy of conservation paradigm by inserting the concept of resources rent, the return to resources owners from the fishery. In this paradigm, fisheries should be run rationally and economically efficient. This rationalisation paradigm focuses on the achievement of an economically rational or efficient fishery.

In contrast to the conservation and rationalisation paradigms, public policy debates often revolve more around human concerns. This paradigm, called social/community paradigm, views the best means to achieve fishery sustainability is through a complex and systematic analysis on community-based means, capable of controlling harvests, making use of appropriate technology, and promoting long-term resilience and diversity. In this paradigm, sustainability is not achieved by only focusing on the conservation of the fish nor maximising the economic rents but rather by preserving the way of life in fishing communities.

SUSTAINABLE FISHERY

A conventional idea of a sustainable fishery is that it is one that is harvested at a sustainable rate, where the fish population does not decline over time because of fishing practices.

Sustainability infisheries combines theoretical disciplines, such as the population dynamics of fisheries, with practical strategies, such as avoiding overfishing through techniques such as individual fishing quotas, curtailing destructive and illegal fishing practices by lobbying for appropriate law and policy, setting up protected areas, restoring collapsed fisheries, incorporating all externalities involved in harvesting marine ecosystems into fishery economics, educating stakeholders and the wider public, and developing independent certification programmes.

Some primary concerns around sustainability are that heavy fishing pressures, such asoverexploitation and growth or recruitment overfishing, will result in the loss of significant potential yield; that stock structure will erode to the point where it loses diversity and resilience to environmental fluctuations;

that ecosystems and their economic infrastructures will cycle between collapse and recovery; with each cycle less productive than its predecessor; and that changes will occur in the trophic balance (fishing down marine food webs).

DEFINING SUSTAINABILITY

The notion of sustainable development is sometimes regarded as an unattainable, even illogical notion because development inevitably depletes and degrades the environment.

Ray Hilborn, of the University of Washington, distinguishes three ways of defining a sustainable fishery.

- Long term constant yield is the idea that undisturbed nature establishes a steady state that changes little over time. Properly done, fishing at up to maximum sustainable yield allows nature to adjust to a new steady state, without compromising future harvests. However, this view is naive, because constancy is not an attribute of marine ecosystems, which dooms this approach. Stock abundance fluctuates naturally, changing the potential yield over short and long term periods.
- Preserving intergenerational equity acknowledges natural fluctuations and regards as unsustainable only practices which damage the genetic structure destroy habitat, or deplete stock levels to the point where rebuilding requires more than a single generation. Providing rebuilding takes only one generation, overfishing may be economically foolish, but it is not unsustainable. This definition is widely accepted.
- Maintaining a biological, social and economic system considers the health of the human ecosystem as well as the marine ecosystem. A fishery which rotates among multiple species can deplete individual stocks and still be sustainable so long as the ecosystem retains its intrinsic integrity. Such a definition might consider as sustainable fishing practices that lead to the reduction and possible extinction of some species.

Social Sustainability

Fisheries and aquaculture are, directly or indirectly, a source of livelihood for over 500 million people, mostly in developing countries. While biodiversity is important, people need food security.

Social sustainability can conflict with biodiversity. A fishery is socially sustainable if the fishery ecosystem maintains the ability to deliver products the society can use. Major species shifts within the ecosystem could be acceptable as long as the flow of such products continues. Humans have been operating such regimes for thousands of years, transforming many ecosystems, depleting or driving to extinction many species. According to Hilborn, the "loss of some species, and indeed transformation of the ecosystem is not incompatible with sustainable harvests. " For example, in recent years, barndoor skates have been caught as bycatch in the western Atlantic. Their numbers have severely declined and they will probably go extinct if these catch rates continue. Even if the barndoor skate goes extinct, changing the ecosystem, there could still be sustainable fishing of other commercial species.

Reconciling Fisheries with Conservation

At the Fourth World Fisheries Congress in 2004, Daniel Pauly asked, "How can fisheries science and conservation biology achieve a reconciliation?", then answered his own question, "By accepting each other's essentials: that fishing should remain a viable occupation; and that aquatic ecosystems and their biodiversity are allowed to persist. "

A relatively new concept is to design relationship farms, which restore the food chain in an area. Once a healthy food chain is re-established, production is so high that the farm can feed the food chain, produce high yields and filter out impurities from feed water and air. An example of this is Veta La Palma. Veta la Palma is located in southern Spain and lies at the centre of the Guadalquivir River's estuary marshes. It was a large cattle ranch that was re-purposed in a thoughtful way, and is one of the largest and most successful examples of relationship farming.

Relationship farming is concept first made popular by Joel Salatin, who created a relationship farm. Salatin's 550-acre (2. 2 km^2) farm is featured prominently.

The basic concept of relationship farming is to put effort into building a healthy food chain, and then the food chain does the hard work.

SUBSISTENCE FISHERS

Commercial and subsistence fishers hunt fish in wild fisheries or farm them in ponds or in cages in the ocean. They are also caught by recreational fishers, kept as pets, raised by fishkeepers, and exhibited in public aquaria. Fish have had a role in culture through the ages, serving as deities, religious symbols, and as the subjects of art, books and movies. Fish are a huge group of animals, represented by over 28 000 different species. They are the oldest as well as the largest group of vertebrates.

They are characterised by being almost exclusively aquatic throughout their lives. They are limbless, but have two sets of paired fins as well as a variety of single fins. They breathe predominantly by using organs known as gills. They are cold-blooded (poikilothermic), which means their body temperature varies with that of their surroundings. Their skin is usually covered with scales. Most temperate freshwater fish are called teleost (' true bone ') fish, which means that they have a bony skeleton. The only fish found in temperate freshwater that do not belong to this group are the sturgeons and the lampreys. Neither of these groups is found commonly in still-waters, although a few species of sturgeons have been introduced into still-waters in recent years. Teleost fish are extremely successful and well adapted to their environment.

FISH S PECIES

There are around 50 species of fish found for at least part of their lives in the freshwaters of the British Isles. This is quite a small number of species considering the diversity of the aquatic habitats available to them and the number of species in similar waters in continental Europe. Although many of the species are native to the British Isles, there are also many that have been successfully introduced over the centuries. Some of these have settled and become well established and are now generally considered ' native ', for example the carp.

Some, however, still raise concerns among conservationists about the potential impact they may have, should they become more widespread; the zander and wels catfish for example, are described as ' non-native ' and are subject to regulation. Fish species have developed and evolved to make the best use of certain habitats. Many of the species present in the British Isles are especially developed to live in rivers. These species are usually unsuited to living in still-waters as they are not adapted to this very different type of habitat. In this chapter we will concentrate on those species already adapted to the still-water, and those that are commonly stocked in still-waters with apparent success, such as the chub.

The barbel has been included, as it is stocked commonly in still-waters, but it should be borne in mind that special conditions must be maintained for this species owing to its requirement for clear, oxygen-rich water. This chapter sets out some detailed information about each species, which should allow fishery managers to understand better the species under their custody. The ecology of lakes, should allow the manager to establish the best conditions for the fish species and assess any other species for likely success in any given situation.

The fish described in this chapter are the following:

Family Cyprinidae

- Barbel
- Bream
- Silver bream
- Common carp
- Chub
- Crucian carp
- Goldfish.

Cyprinidae

The barbel is a large river fish with a long, streamlined, slightly flattened body. The juveniles are an overall brown colour but as they get larger they develop a dark brown back with golden bronze flanks and a creamy white underbelly. Adults generally grow to around 70 cm, but can get up to 90 cm and weigh in excess of 6 kg (the British rod-caught record 2008 is 9. 837 kg). The barbel is native to the southern UK and mainland Europe. It is widespread and relatively common.

The barbel will generally mature after 3-5 years. Spawning takes place once a year in the late spring, usually May-July. Barbel spawn in small groups among gravel and stones in fast-flowing rivers. The adhesive eggs stay within the gravel bed until they hatch 10-15 days later. The fry tend to congregate in slow flowing areas where they feed on zooplankton. As they mature and get larger they tend to move out into the main flow of the river, feeding on invertebrates and even small fish in among clean gravel and associated vegetation.

The barbel is a very popular coarse fish, being one of the most sought-after river fish. They are a hard fighting fish that can grow very large, attracting specialist anglers.

Barbel have increasingly been stocked in still-waters over several years. They appear to grow and survive reasonably well, but there is little evidence that they can successfully spawn in most still-waters. It is vital that any stocking is with barbel obtained from a reputable fish farm. Barbel can sometimes be introduced from fish poached from the wild. Not only is this very harmful to the wild populations but evidence suggests that these fish find it very difficult to make the transition from rivers to still-waters. Generally barbel should not be stocked in shallow still-waters with a predominance of bottom feeders such as carp.

Perch (Perca fluviatilis)

The perch is a beautiful, deep-bodied fish with four to six dark bands over olive-green flanks. The abdomen is creamy silver with prominent red pelvic and anal fins. It can grow to 20 – 40 cm and weigh 1 – 2. 5 kg (the British rod-caught record 2008 is 2. 523 kg). The perch is native to the UK and Europe and is now widespread and common in the UK, northern Europe and Asia. It is a popular food fish in many countries; in England it used to be canned and sold as ' perchines '.

Perch generally mature after 3 – 4 years and live to around 10 years. They spawn in early spring, usually March to May. Perch tend to spawn in the margins, producing long strings of eggs which lie wrapped around submerged vegetation. The eggs hatch after 10 days and the juveniles feed on small invertebrates as well as fish fry, including other perch fry. As they get older they start to feed on larger invertebrates and begin to feed actively on other fish. Small perch shoal in groups but as they get bigger they do become more solitary, with large individuals tending to move around in small shoals of just a few individuals.

The perch is a popular fish for coarse fisherman, being one of the most attractive of the freshwater fish species. It responds well to anglers ' baits and is readily caught. As a predator it is a useful fish to control the populations of other fish species. However, they can mature at a very small size (10 – 15cm) and in certain circumstances they can create an abundance of small perch. It appears that perch only grow to a large size where there is a plentiful supply of small prey fish.

Zander (Sander lucioperca)

The zander is a large predatory species with a long, thick body. The back is a dark brown – green with obvious dark brown vertical stripes running to the middle of the body. It grows quite large, reaching average sizes of 50 – 70cm, occasionally much bigger to a maximum of 120 cm (the British rod-caught record 2008 is 8. 7 kg). The zander is common throughout Europe. It was deliberately introduced into England (the River Great Ouse) in 1963 and has spread widely since.

Zander generally mature after 3 – 5 years. Spawning takes place from April to June in among marginal vegetation. Pairs of fish spawn together at dawn and lay sticky eggs on vegetation and stones on the bottom. Unusually, both parents will guard the eggs until they hatch 5 – 10 days after spawning. The fry start on a diet of invertebrates but move quickly on to eat mainly of other fish. The zander is a hunter, patrolling open waters and chasing down small prey fish. Occasionally it appears to hunt as a group, although it mainly hunts as an individual.

The zander is a very popular fish with anglers, being a fast-growing, large and aggressive fish. It survives well in larger, clear still-waters and may be a useful addition to a specimen water as it prefers small fish as prey. Licences are required to stock this species.

Gasterosteidae

Three-s pined s tickleback *(Gasterosteus aculeatus)*

The stickleback is one of the most common and recognisable species in freshwaters. It is a small, thin-bodied fish with an olive-brown back and silvery sides and underside. The body is covered in large bony plates and there are three obvious strong spines on the back. The stickleback is a very small fish that rarely exceeds 10 cm in length (the British rod -caught record 2008 is 7 g (0. 007 kg)). Sticklebacks are widely distributed and common throughout Europe. Sticklebacks mature after 1 – 2 years. Spawning takes place in March

– June. They are unusual in that the male builds a nest of fibrous plant material and attracts females to spawn within it. The male also develops a vivid red throat to help attract a mate. The eggs take 10 – 20 days to hatch and are guarded constantly by the male. This guarding behaviour continues until the fry are a few weeks old. The fry feed on small invertebrates. As they get older they continue to feed on invertebrates but may extend their diet to include fish fry. Sticklebacks tend to aggregate into small shoals, only becoming solitary and territorial during spawning times.

The stickleback is not a target species for anglers. It is common in most still-waters and will almost certainly appear without help.

Nine-s pined s tickleback *(Pungitius p ungitius)*

This is a rare species. It is similar in most respects to the three-spined stickleback, except that it has 7 – 12 spines running along its back. It also goes dark with a black throat when in breeding condition.

Common Bream (Abramis brama)

The common bream is a large, bottom-feeding species with a deep thin body. When young it is a very silvery fish, but as it matures it becomes darker and more golden brown to green. It can grow to nearly a metre in length and weigh up to 8 kg (the British rod-caught record 2008 is 8. 329 kg). The common bream is native to the UK and central Europe where it is widespread and common. It is a popular food fish in eastern Europe.

Bream generally mature after 4 years. Spawning takes place in May and June in dense vegetation in shallow water. The males develop white bumps (tubercles) over the body surface and become territorial. The females spawn at night in large groups and their sticky eggs attach to the vegetation. After 5 – 10 days the eggs hatch.

The fry start by feeding on zooplankton such as rotifers and then larger zooplankton such as daphnia. They quickly move on to their normal adult diet of benthic invertebrates, preferentially worms, molluscs and insect larvae such as chironomid midge larvae. The adults feed by using their extendable tubular mouth to root around in the pond bottom, taking large mouthfuls and ' winnowing ' out the food items. They are strongly shoaling fish from larval fry to late adulthood.

Some of the very large individuals appear to be solitary or in small shoals. The common bream is one of the most popular of the coarse fish. It responds well to anglers ' baits and is readily caught. It grows quickly to a good size and survives well in many different conditions. As a bottom feeder it will disturb the bottom of the pond, causing suspended solids to cloud the water column. They are good in a mixed silver-fish population. The common bream should not be mixed with large numbers of carp as the species compete for the same food source.

The silver bream is a small mid-water species, with a deep, compressed body. When young it is a very silvery fish but as it matures its back becomes a darker olive/grey, although it still retains a very silvery look. Its maximum size is around 20 – 25 cm in length (the British rod-caught record 2008 is 0. 425 kg). The silver bream is native to England, central and northern Europe, where it widespread and reasonably common. In England it is generally restricted to the southeast but is becoming more widespread.

Silver bream mature at around 3 – 4 years old. Spawning takes place in May – July in dense vegetation in shallow water. They spawn in groups early in the morning, scattering their sticky eggs among the submerged weeds. The eggs hatch in 4 – 8 days and begin feeding on small planktonic invertebrates, moving to larger invertebrates and plant material as they grow. Silver bream form strong shoals with individuals of a similar size and age throughout their lives. The silver bream is not a popular species for anglers, largely because it is commonly mistaken for a small common bream. Indeed it is very similar to a common bream of the same size, except the silver bream is slightly thicker in the body and has larger scales. They do make useful additions to a lake as they feed in a very different manner.

The carp is a large, bottom-feeding species with a thickset rounded body. They are usually a dark olive brown on the back, lightening to a light yellowish brown on the belly. It can grow to very large sizes, sometimes exceeding 1 m in length (the British rod-caught record 2008 is 26. 9 kg). The carp is native to eastern Europe and Asia, but has now achieved worldwide distribution. It is very widespread and common in the UK, where it was first introduced in the nineteenth century.

Carp mature after 2 – 3 years. Spawning takes place in June to July in dense vegetation on the margins of the water body. The males develop small white lumps (tubercles) on the flanks and head. The females spawn at dawn and the sticky eggs attach to the weed. The eggs will hatch after 5 – 7 days and the fry start feeding on zooplankton, particularly rotifers. They quickly move to feeding on the bottom-dwelling invertebrates that make up the bulk of their diet. Their mouth is well designed for this feeding technique, being protrusible. However, carp are a very adaptable species and they can often be seen feeding on invertebrates on marginal vegetation and even off the surface. They tend to move around in small shoals of similar-sized individuals.

The carp is probably the most popular species for anglers on still-waters. It grows quickly to very large sizes and responds well to a variety of anglers ' baits. It is a very adaptable species so can survive and thrive in many different conditions. As a bottom feeder, it will disturb the sediment on the bottom of the water body, causing suspended solids to cloud the water. Carp have been bred for food over many centuries and it is one of the main global species for aquaculture in ponds. This has led to several different varieties of carp being commonly found. The most obvious of these varieties is the common carp, which is the fully scaled type. Mirror carp were bred to make it easier to remove the scales before cooking. Ornamental varieties are also common, from full-colour varieties, known as koi, to slightly gold-tinged cross breeds known as ghost carp. These are all the same species of fish, however.

The chub is a large, mid-water and surface-feeding fish with a rounded, streamlined body. When young it is a very silvery fish with an olive green back; as it gets larger its upper body darkens to a steel grey colour but retains the silvery flanks and belly. It is a large fish, attaining sizes of between 40 and 50 cm, rarely getting to up to 80 cm (the British rod-caught record 2008 is 3. 9 kg). The chub is native to the UK and central Europe and is widespread and common.

It is caught for food in some parts of Europe. Chub generally mature after 3 – 4 years. Spawning takes place during April – June among clean gravel and stones in flowing water. They spawn in the early morning in small groups of one female with several males in attendance. The eggs hatch within 6 – 8 days and start to feed on benthic invertebrates. As they grow older their diet changes to include larger invertebrates, fruit, vegetation and even small fish. When young, chub move around in shoals but as they mature they become much more solitary and even appear to be somewhat territorial. In rivers they migrate significant distances to feeding and spawning areas. Little is known about their activities in still-waters.

The chub is a popular fish among anglers because it is a large, hard-fighting fish that responds well to anglers ' baits. It grows quickly to a good size and appears to do very well in still-waters. Current evidence suggests that it does not successfully spawn in most still -waters, although it does spawn in large, clear-water lakes.

The crucian carp is a medium-sized, deep-bodied fish with marked lateral compression. Young fish are a golden bronze colour, which turns darker as they get older, with mature specimens being olive brown on the back through deep bronze on the flanks to a golden yellow on the belly. Adults can attain 20 – 30 cm in length and 1. 5 kg in weight, although many crucian carp populations remain ' stunted ' with the mature adults rarely getting more than 15 – 20 cm and 0. 5 kg (the British rod-caught record 2008 is 2. 01 kg). The crucian carp is native to the UK, Europe and central Asia.

There is some argument about whether it is a native to the UK and western Europe, but there seems little evidence to suggest it is introduced into these areas, where it is widespread. Crucian carp generally mature after 2 – 3 years. Spawning takes place in May – June in shallow water, in thick vegetation. The fish shed sticky eggs onto the vegetation where they take between 5 and 7 days to hatch, depending on the vegetation. Once hatched, and after the yolk sac has been used, the fry tend to form loose shoals and feed on small zooplankton. As they grow they start to feed mainly on invertebrates, such as molluscs and worms on the bottom of the pond, as well as plant material. Both the juveniles and the adults tend to live in loose shoals. They tend to occupy the mid-water and bottom of still-waters. Crucian carp prefer still-waters and particularly thrive in rich, lowland still-waters with abundant vegetation. Crucian carp have a remarkable ability to survive very low dissolved oxygen situations for long periods, leading to situations where they are often the only fish species to survive. Here they form healthy populations dominated by small ' stunted ' individuals. In larger waters they can grow much larger and seem to adopt a slightly different body shape.

This is a very popular angling species and is readily caught. It survives well in well -managed fisheries dominated by silver fish, but it is generally unsuitable for stocking in waters dominated by common carp. Crucians readily hybridise with goldfish and common carp. They are then displaced by the vigorous hybrid. They are also threatened by the non-native parasitic tapeworm *Bothriocephalus acheilognathi*.

The goldfish is a medium – sized mid-water species with a relatively deep and laterally compressed body. Wild goldfish are a deep brown colour throughout their lives, although there are many colour variants from deep orange through to almost white. Adults can get to around 30 cm and 1 kg in size. The goldfish is native to Asia, but owing to its ornamental value it has achieved worldwide distribution, being found on most continents. It is an introduced species in the UK.

Goldfish generally mature after 2 – 3 years. Spawning takes place during June and July in thick weed in the shallow margins. The eggs hatch in 4 – 7 days, depending on the water temperature, and the young feed on zooplankton. As they grow, the diet changes to small invertebrates and plant material. Goldfish tend to form loose shoals and prefer shallow, rich ponds with abundant vegetation. They are a hardy species, however, and appear to thrive in most still-water conditions.

The goldfish has become increasingly common in still-waters in the UK and Europe. It is apparently popular with some fishermen.

Goldfish readily hybridise with common carp and crucian carp, producing vigorous offspring that can successfully outcompete the original parent stock. Stocking with goldfish or hybrid strains is advised against.

The grass carp is a large mid-water feeding species with a solid, rounded body. It has a dark blue-black back, shading to silver on the belly. It can grow quite large, with adults reaching up to 100 cm in length (the British rod-caught record 2008 is 20. 185 kg). Grass carp are originally from Asia, but have been widely distributed around the world as a method of weed control. They were introduced into England in the 1960s to control weed growth in some still-waters. Since then they have become widely distributed, but because of their specialised breeding requirements there appear to be no breeding populations. Grass carp mature after 3 – 4 years. Spawning takes place in fast-flowing water in the headwaters of large rivers. Temperatures must be around 23 – 25 ° C. The eggs are unusual in that they are pelagic, being carried down river in mid-water until they hatch in the slower flows downstream. The fry start by feeding on small invertebrates, but as they get older they begin to feed on aquatic plants. Grass carp form loose shoals throughout their lives.

Grass carp are not particularly popular among anglers as they are generally difficult to catch and do not fight as hard as their size might suggest. They are not very effective in controlling weed growth as they do not feed until the waters are quite warm.

The gudgeon is a small, rounded, bottom-dwelling fish. Its back is a dark olive green to brown with lighter flanks and a silvery belly. It has dark markings along the flank and upper body. It is a very small fish, rarely reaching over 15 cm although it can grow to about 20 cm. (the British rod-caught record 2008 is 0. 141 kg). Gudgeon are widely distributed throughout Europe and Asia and they are native to the UK.

Gudgeon mature at between 2 and 3 years old. They will spawn on stony or gravely bottoms, or among weed, and prefer flowing water, although they will spawn successfully in still-waters. They spawn during May or June and the eggs hatch in around 15 days. They feed on insects, molluscs, crustaceans and occasionally plant material. They are generally a solitary fish, but sometimes form into small shoals of perhaps a dozen individuals.

The gudgeon is not a popular target species for anglers, but it does do well in some still -waters where it probably makes an important contribution to the health of the lake.

The minnow is a small mid-water species with a long rounded body. It has a brown-gold body with numerous brown and black spots down the flanks usually joining to form an obvious black line along the body. It is a small fish, rarely exceeding 10 cm in length (the British rod-caught record 2008 is 0. 135 kg). The minnow is a native to the UK, Europe and Asia where it is very widespread and common.

Minnows mature after 1 – 2 years. Spawning takes place in June or July, usually over clean stones or gravels. The males particularly become more colourful during spawning, with a scarlet belly and black throat. The eggs take 5 – 10 days to hatch and immediately start feeding on small invertebrates. As they grow they continue to feed on a variety of invertebrates. Minnows generally aggregate into small, loose shoals.

The minnow is not a target fish for anglers, although it can be important as a bait fish. It does well in lakes and ponds and can be a very useful addition to the lake ecology, particularly if predatory fish are present. The orfe is a large, mid-water and surface-feeding fish with a rounded, streamlined body. When young it is a very silvery fish, but as it gets larger its upper body darkens to a brownish grey colour. Coloured ornamental varieties are common, particularly golden and blue orfe. It is a medium-sized fish, attaining sizes of between 30 and 40 cm, although it can occasionally get up to 80 cm. (the British rod-caught record 2008 is 3. 7 kg). The orfe is native to central and eastern Europe and parts of Asia, where it is widespread and common. It is caught for food in some parts of Russia. It was introduced into England and is now widespread and common.

Orfe generally mature after 3 – 4 years. Spawning takes place during April and May, among weeds and clean stones in shallow water. The eggs hatch within 15 – 20 days and start to feed on benthic invertebrates. As they grow older their diet changes to include larger invertebrates, although they will take fruit, vegetation and even small fish. Orfe move around in shoals of similar-sized and aged individuals. They are found equally in rivers and still-waters.

The orfe is a popular fish among anglers because it is a good-sized, hard fighting fish that responds well to anglers ' baits. It grows quickly, to a good size and it appears to do very well in still-waters. The roach is medium-sized, mid-water fish with a relatively deep body. The young fish are silvery, but they quickly develop a dark blue – green back and light underside. Adults generally get to 20 – 40 cm and between 1 and 2 kg (the British rod-caught record 2008 is 1. 899 kg). The roach is native to the UK and Europe, where it is very widespread and common throughout the lowland catchments.

Roach generally mature after 2 or 3 years. Spawning takes place once a year in the middle of spring, usually April – June. The roach spawn at dawn, shedding their sticky eggs on dense, submerged vegetation in the shallow areas of the lake.

The eggs will hatch in 5 – 10 days. Once the yolk sac is used, the fry will feed on zooplankton and other small invertebrates. The juveniles prefer to live in large shoals in the shallow, marginal areas with plenty of vegetation, which provide cover and food. As they mature and get larger they move out into open water, shoaling in smaller groups of similar-sized individuals. These larger individuals feed on larger benthic invertebrates and pondweed, particularly attached algae (periphyton).

They tend to occupy the mid-water to bottom of the pond. The roach is one of the most popular coarse fish in the UK. It responds well to anglers ' bait and is readily caught. It grows steadily and survives well in many different conditions. It is an excellent species in any mixed fishery as it is very adaptable. In the right conditions, it can quickly overpopulate with a very large population of small fish.

The rudd is a medium-sized, top-feeding species, with a deep, laterally compressed body. It has a dark greenish-brown back, grading through to a silvery underside. The flanks are golden brown. The iris is a deep gold with a red fleck on the upper side. The fins are normally a bright red. It can grow to 20 – 25 cm and 0. 5 – 1 kg (the British rod-caught record 2008 is 2. 1 kg). The rudd is native to the UK and central Europe, where it is very widespread and common.

Rudd generally mature at 2 – 4 years and live to 10 – 14 years. They spawn in late spring, usually from April to July. Rudd spawn by shedding their eggs on submerged vegetation around the margins. The eggs hatch in around 5 – 10 days and the fry stay within the protection of the vegetation, feeding on small zooplankton. As they grow they form compact shoals, feeding on the surface and in midwater. They prefer zooplankton, weed-dwelling invertebrates and terrestrial insects on the surface. Rudd appear always to spend their time in fairly compact shoals, no matter how large they get.

The rudd is a popular fish to catch, being attractive and readily taking bait, which is particularly important if other species are more reluctant to feed on anglers ' bait. It grows steadily but rarely attains large sizes. It is a very good fish for a mixed fishery, as it does not compete directly with any other fish species.

The tench is a large, bottom-feeding species, with a thickset, slightly laterally compressed body. The tench is a beautiful olive to dark green throughout its life, with small orange to red eyes. Adults generally grow slowly to a maximum length of 60 – 70 cm and can weigh up to 7 kg. (the British rod-caught record 2008 is 6. 9 kg). The tench is native to the UK and Europe, although absent from the northern parts of both. The species has been widely introduced into Asia and North America.

Tench generally mature at 3 – 4 years old. Owing to their slow growth, they may only be 15 cm at this age. Spawning takes place once a year in spring, usually between May and July. They spawn among aquatic plants in shallow waters, to which their sticky eggs attach. The eggs will hatch in 4 – 5 days, and once the yolk sac is used the fry will feed on zooplank-ton, before switching to grazing among the bottom of water plants on a variety of aquatic invertebrates and attached algae. Tench prefer still-waters, or very slow flowing rivers with an abundance of macrophytes. The juveniles may form loose shoals, but as they mature they become more solitary, almost appearing to have preferred territories.

The tench is one of the most popular and easily recognisable of the coarse fish in the UK and Europe. Its popularity relates to its size and its beauty. It is generally a difficult fish to catch on rod and line. It is quite adaptable to a mixed fishery, but does not thrive in waters without good communities of aquatic macrophytes.

The pike is a large predatory fish, with a slim streamlined body and a large head with an enormous mouth full of teeth. It has a dark green, mottled back, grading through light green flanks to a creamy yellow underbelly. It is a very large fish, routinely growing to 50 – 100 cm and occasionally 140 cm and 30 kg (the British rod-caught record 2008 is 21. 2 kg). The pike is widespread and abundant throughout the Northern Hemisphere in Europe, Asia and North America. It is a popular food fish wherever it is found.

Pike generally mature after 3 – 4 years, the males earlier and smaller than the females. They spawn among weed during February – May. The eggs attach to the weed and hatch after 10 – 15 days. When they hatch they start by feeding on invertebrates but quickly turn to small fish and then onto larger prey, including other vertebrates such as ducks and water rats. They are very partial to other pike and are, unsurprisingly, very solitary and territorial. It appears that females grow much larger than males, and it is said that any pike over 5 kg will be a female. They are an ambush predator, laying in wait for their prey and only striking at the very last minute when the prey is within one to two body lengths. The pike is one of the most popular freshwater fish in the UK. It grows exceptionally quickly to a large size and is a ferocious fighter, particular in the first moments as it attacks the bait.

Percidae

The ruffe is a small fish with an obvious spiny dorsal fin, large eyes and mouth. Its back is a dark green-brown colour with irregular dark blotches, whereas its underside is pale yellow to cream in colour. It usually grows to around 15 cm but can exceptionally grow to 25 cm (the British rod-caught record 2008 is 0. 142 kg). The ruffe is native to much of Europe but was confined to eastern England, although its range has spread considerably in recent decades.

The ruffe matures after 1 to 2 years. They spawn in shallow water among stones and vegetation, mainly in April to May, although they probably spawn several times in a year. Initially the fry grow quickly on a diet of small invertebrates before moving on to larger invertebrates and other food such as fish eggs and young fry. The ruffe is a shoaling fish, moving around in groups throughout its life. The ruffe is not a popular fish for anglers, being such a small size and attacking baits destined for other larger target species. It has numerous alternative common names such as pope, tommy and daddie.

EVALUATION OF SUSTAINABILITY FISHERIES

It is argued that sustainable fisheries should be defined widely. While there is general agreement that resource conservation is necessary for fishery sustainability, the concept of fishery sustainability must involve more options and objectives, including other human concerns instead of rational objectives such as MEY and OSY. Pitcher and Preikshot have also mentioned that conventional stock assessment (conservation paradigm) relates to the ecological, or occasionally the economic rationale paradigm, and yet fisheries in reality are a multi-disciplinary human endeavor that has social and ethical implications, besides policy implications. Evaluation of sustainability in all of these disciplines is required for objective decision-making. In other words, while the balancing of present and future catches is important, there is more to a healthy future than simply a large fish stock.

It is also important to pay attention to sustaining the process underlying the fishery. Moreover, Charles mentioned that what is sometimes missing from the discussions of fisheries sustainability is the attention to the state of human system. This is where the concept of sustainable development becomes important.

A well-known definition of sustainable development is the one given by WCED as the development that meets the needs of present generation without compromising the ability of future generations to meet their own needs. Following this definition, there is wide recognition of the need to view sustainability broadly, in an integrated manner that includes ecological, economic, social and institutional aspects of the full system (*i. e.* ,, fishery system).

IDEA OF A SUSTAINABLE FISHERY

Fisheries have been known to be an important activity throughout the world, and produce more than 100 million tons of fish and fishery products and contribute to human welfare by providing a livelihood to about 200 million people, and protein supply for a billion people. However, with declining stocks as well as several evidences related to fisheries, sustainability issue became very important and has been discussed as the central topic in fishery sciences and industries.

This condition is mainly encouraged by the unfortunate reality that many fisheries are in a state of crisis, and some of them demanding urgent attention. At global levels, fishing industry is a highly adaptive, market-driven and dynamically internationalised sector within the world economy. Its pressure on resources is still increasing, owing to a persistent worldwide upward trend in fish consumption, in concert with human population growth, especially in coastal zones. Global efforts are increasing and have limited the capacity of individual governments to control the pressure over fishing. This problem was associated with a variety of environmental and ecosystem problems including wastage from discards, loss of critical habitats, impact on endangered species, etc. .

BIOLOGICAL CRISIS OF FISHERY MANAGEMENT

From this point, fishery sciences itself are progressively switching their attention from single species to ecosystem approach, from micro to macro perspectives, increasing the need for measuring the impact of fishing on natural and man-made systems. Consequently, Cochrane argued that the problems currently experienced in fishery management throughout the world occur in four realms, namely biological, ecological, economic and social crises.

Biological crisis of fishery management started during the last decade after awareness grew about the alarming status of fishery resources. Ludwig for example, drew attention to collapses of fish stocks such as those of the Pacific salmon, the Californian sardine, and the Peruvian anchovy. In 1994, FAO showed, in an analysis, of global fish landings, that there had been reduction in the annual growth in landings in 1980s, and that in 1990 there had been 3 per cent reduction in the global annual catch from the previous year. Garcia and Newton analysed that this trend continued over the next few years and in between 1990 and 1992, global landings fell by an average of 1. 5 per cent per year.

In an economic perspective, fisheries actually exist to meet social and economic demands and one would expect to find that the impact on the resources has resulted in a measurable social and/or economic benefit. Unfortunately, some evidences suggest that the expected benefits have not been in the form of economic gain. Christy estimated that the gross revenues from the total global marine landing in 1989 was US$ 70 billion, giving a deficit balance since the global operating costs were estimated at a level of US$ 92 billion. With an annual capital costs of US$ 32 billion per year, the total deficit in global fisheries was estimated to be US$ 54 billion per year.

At a country level, Japan for example, a deficit gross cash flow was also founded for cause of purse seine fisheries. In this type of fisheries, the volume of production was estimated to be lower than its MYS, OSY as well as MEY level. However, due to its high operating cost and low price of fish caught by the purse seine, it resulted in a deficit cash flow. Matsuda called this phenomenon the poverty in large catch. Nevertheless, the recent study undertaken by Tietze shows that in some fisheries, they perform a positive gross cash flow in their global operation. Out of 108 types of vessels from 15 countries throughout the world, 97 per cent had a positive gross cash flow and fully recovered their operating costs. When also considering the annual cost of capital, 85 per cent showed a net profit after deducting the cost of depreciation and interests. However, in terms of sustainability issues, some symptoms of economic crisis in some fisheries should be considered in order to come up with the solution for economic sustainability in the system.

From a social point of view, fisheries are rarely seen as simple tools for generating economic returns. The role of fisheries as a source of employment, particularly in rural or more remote areas, has also widely been given high priority. Moreover, McGoodwin elaborated that fisheries actually are a human phenomenon which essentially are places where human activities are linked with marine ecosystems and renewable resources. Human fishing activities defined the attribute of fishery, since without it there would only be an aquatic realm where various marine species live. Therefore, it has been strongly argued that clearly the fisheries are much more than geographic regions, fishing methods, types of fishing gear, fish species, or economic-domains- it is something more human.

In this regard, social aspects in fisheries would be very important to be understood. One of the major features of fisheries in recent decades, related to the social crisis, has been the introduction of modern fishing technologies and also the increasing globalisation of trade affecting the fishing communities. Modernisation in fisheries has two faces of which actually one contributed to the welfare of fishing communities, while the other has led to the social problem related to the depletion of fishery resources. According to Crean and Symes, in the North Atlantic and Mediterranean, there has been a decline in the quality of life and standard of living among many fishing communities, as also experienced in some parts of South East Asia. This symptom of social crisis in fisheries should seriously be taken up when we consider the sustainability in this sector.

We argue that the need for measuring and evaluating the sustainability of fishery activities in a system perspective has acquired its highest importance and should be undertaken at various levels involving all aspects in the fishery system. As also strongly argued by Charles that critical concern about sustainability arises not only in terms of catch level or even biomass level, but in all aspects of the fishery: from the ecosystem, to the social and economic structure, to the fishing communities and management institutions, as well as fish stocks themselves. Moreover, the pursuit of sustainable fisheries is best seen as requiring, more than keeping the catch of fish, to keep a level that is not too large. Instead, sustainability can be viewed comprehensively as the maintaining or enhancing of four key components, namely ecological, socioeconomic, community and institutional sustainability indicators.

There have been some methods for analysing and evaluating fishery sustainability, such as RapFish method, FAO Code of Conduct Compliance, and International Instrument Compliance. RapFish is a rapid appraisal technique designed to allow an objective, transparent, multi-disciplinary evaluation. It uses multidimensional scaling, and ordination method to appraise the relative sustainability of fisheries. RapFish relies on a defined-scoring of a number of attributes in five dimensions, namely of ecological, economic, social, technological and ethical. As its characteristics of a rapid appraisal, this method has a slightly top-down approach by employing the defined-score and criteria for the fishery system evaluation. The application of RapFish can be seen, for example, in the Gulf of Maine fisheries, in the North Atlantic Fisheries and recently in the case of Jakarta coastal fisheries, Indonesia.

Similar to RapFish, compliance with the FAO Code of Conduct required multidimensional scaling approach. In this case, four different sets of attributes articulate the clauses in the Code. This evaluation method reflects national management intentions and management practices for each fishery system that is assessed. The application of this method has been carried out in Australian fisheries, Australian multi-species trawl fishery, and Gulf of Maine fisheries. International Instrument Compliance method focuses on measuring either qualitatively or quantitatively the level of compliance with the fisheries management provisions contained within an instrument. In this method, criterion scores ranged from 0 (low compliance) to 3 (high compliance) and the total criteria were limited to approximately six per instrument. An example of this method is presented by Chuenpagdee and Alder in the case of Gulf of Maine fisheries.

It could be said that the methods rather focus on relatively static approach from the top. The fishery in question is just used like the static object which will be evaluated using such methods. They do not involve stakeholders to evaluate their fishery. By these methods, a straight forward analysis could be done but should be treated as preliminary results only. We then argue that it is necessary to confirm dynamic complex systems of the fishery by using a bottom-up approach.

FISH FARM PLANNING SYSTEM

Fish culture is one form of aquaculture. Rearing fish in ponds is an ancient practice which began as long ago as 2700 B. C. in China. However, there is a lot of difference between the ancient and the modem pond fish farming practices. As a result of many research studies allover the world on fish farming, the scientists have created a system of knowledge that can be utilised effectively by the modem fish farmers.

Fish farming cannot be carried out in isolation; it should fit into the other farming systems and ultimately in to the ecological system. In the ecological system there exists a continuous cycle of organic matter and water. The planning should be such that the fish farm become a part of the ecological system of the place.

Vegetable and other garden products go to the fish pond while the mud excavated and the sediments formed later from the pond to the garden; the irrigation water from the farm goes to the pond while the surrounding farm area is benefited from the seepage of the water from the pond; the farm waste and animal dung from the farm go to the fish pond as manure while the pond provides fish meal, aquatic insects and plants; at the time of water scarcity water from the pond can be used for vegetable growing and for other essential usages; weeds, grasses and crop wastes from the agriculture farm and the fish farm go to the cattle as fodder while the insects and other aquatic organisms become the feed for poultry and pigs. These are only few examples of how the fish pond fits in with the agricultural system and the ecological system.

Anyone who gets ready to start a fish farm should keep in mind the following points. These points are so important that they are on "Construction of Fish Ponds". For the benefit of the reader of this booklet these points are expressed here in the form of questions for which the prospective fish fanner should find answers and based on the answers he will be able to make a proper decision whether to have a fish farm or not. The answers also will provide him clear ideas to plan for the fish farm.

- Is the soil able to hold water for a fish pond?
- Is there an adequate supply of water for a pond?
- Is the land in a good shape for a fish pond?
- Is the pond area close to your home?
- Who owns the land where the pond will be built?
- Are there enough people to help to build and harvest the pond?
- Can the equipment for building a pond be built, borrowed, or bought?
- Is there a market place nearby?
- Are there roads from the pond area to a market place?
- Are the roads passable even in the rainy season?
- Is there a vehicle available for transportation, if necessary ?
- If there is no market nearby, or if it is hard to get to the market, can the fish be kept by drying, smoking, or salting?
- Is there enough food for the pond fish?
- Are there fertilizers available?
- Do the people in the are alike fish? Do they eat fresh water fish?
- Can the people in the area afford to buy the fish produced in the pond?

If the farmer finds "yes" to most of these questions he has a fairly good chance for success in his fish farming enterprise.

Various aspects of the fish farm planning is discussed in general so that anyone who wants to start a fish farming can find this booklet as a useful guide. This booklet has details for the most sophisticated and commercial fish farmers as well as for the home level fish growers. Hence, it is up to the individual how much of the details given in this booklet are applicable to himself or herself.

PLANNING FOR SITE

Before the construction of the pond the farmer should have a good look at the land available to him to choose the place or places where the ponds can be constructed. Finding a suitable place is the most important and the fist step in the planning of the fish pond. Fish rearing is a form of land use like land use in agriculture. If the site is well chosen the pond will be more productive. Often poor agricultural land can be converted into very good fish ponds. However, the better the soil the greater the possibility of making a good fish pond at lesser expense. There are mainly three factors that contribute to make a good site for a fish pond water supply, soil and topography. These are further explained in detail.

Water Supply

Fish live in water and hence the most important factor for the site selection is the availability of water. If natural water supply is not existing try to find some underground water sources or whether water can be brought in from outside through pipes, channels etc. It is also possible to store rain water into tanks in which the fish can be grown.

The sources of water for the fish pond are mentioned here.

- *Rainfall:* Some ponds rely on rainfall alone and they are called "sky ponds".
- *Run-of:* Some ponds get their water from the run-off from the surrounding areas during the rainy season.
- *Natural waters:* The word natural waters is referred to the water available from the lakes, rivers, streams and springs.
- *Springs:* Some ponds will have active springs inside, filling the pond.
- *Wells:* Many ponds are fed with water from the open wells or tube wells.

However, there can be problems also with the water supply. For example, there can be flooding during the rainy season which will affect adversely the fish pond system. Hence, the places that are prone to flood should not be chosen; even sporadically flooded areas should be avoided for fish ponds unless there are possibilities of protecting the pond area artificially. If the water is taken from the river or stream that are prone to flood there should be an inbuilt mechanism to control the inflow of water into the pond.

Quality of Water

The quality of water is checked by the following methods.

- Look at the water to observe the colour and visible suspended particles, check for any obnoxious smell and taste to check whether the water is salty or having any other taste.
- Check whether there are any people residing or factories situated in the upstream which may pollute the water coming to the pond.
- Make sure also that there are no families or people who depend on the down stream water for drinking and domestic usages.

If the water is very clear the farmer can apply fertilizers in the pond; if the water is muddy he has to allow the sediments to settle down before he uses the water in the pond. He has to build special silt catching structures through which water has to pass before it enters the pond. If the water is greenish in colour probably it has a lot of fish feed in it. If the water is dark, smelly brown it may be acidic and the farmer has to add sufficient lime to neutralise it and add gypsum if the water is alkaline.

Soil

Soil of the area is the most important factor that has to be taken into consideration. It should be able to hold water. Besides the water holding capacity the soil should be rich in nutrients/minerals. Clay soil holds water well and hence, check the soil for the clay content by feeling a sample of it in the hand.

If the soil is sticky and can be formed into shape, it means the soil has high clay content; if the soil is gritty, it is sandy and if it is smooth and silky, it is a silty soil. Clay soil is best for the pond construction. However a poor water holding soil also can be treated at the base of the pond with lot of organic manure and other biomass materials to improve its water holding capacity.

Soils that contain lot of plant nutrients (minerals and elements) are better for the pond construction because the natural growth of micro and macro organisms will be enhanced. In other words, a nutrient rich soil will provide more feed to the fish. At the same time acid and alkali forming nutrients should not be predominant in the soil. One indication of the nutrient status of the soil is to check whether the crops grow well in the soil. If it grows the soil is better for fish pond also. Fertile soils in the pond encourages phytoplankton growth which in turn will allow the growth of macro fauna.

Topography

Topography refers to the shape of the land whether fiat, slope, hilly, upland, lowland, undulated, etc. Topography of the land decides the kind of the pond that can be built. Ponds can be built in any place: in the valleys, on the fiat ground; they can be built in any shape: square or rectangular or uneven shape; they can be large or small. All these are determined by the topography as well as by the farmers requirements.

The best topography for the fish pond construction is the one which allows the farmer to fill the pond and drain the water from it using gravity. Ponds built on the fiat land will have to be made sloping to one side at the bottom to drain out completely using a pump or some other water lifting devices to empty.

Therefore, the best place for the pond construction is a place with slight slope and plenty of water supply. The farmer should look for a topography that makes the fish farming as easy and successful as possible. However, if water is available even on the poor topographical land also fish pond can be constructed.

FISH FARM OPERATIONS

Once the farmer has found out the place for the pond the next step is to decide what kind of fish culture is possible with the space available. He must also decide whether his resources will allow him for starting a fish farm. This planning is necessary to determine the number of fish ponds to be constructed and the kind of fish culture one wants to take up.

A range of ideas are presented in this section of the booklet concerning the kinds of fish farm operations whether raising fish or breeding fish, whether raising carps, tilapia or catfish, whether prawns or shrimps, etc. He also gets an idea on the type of ponds small or big, grow in one pond or in more ponds, advantages of small and big ponds, advantages of mixing or separating different breeds of fish and sexes, etc. A farmer who prepares himself to start fish farming should keep in mind that a certain percentage of fish introduced into the pond will die however much we try to take care of them. In the nature many fishes never reach their adult size because they are eaten by other animals or big fishes or they die of diseases. "Survival of the fittest" is the law of nature. In the artificial rearing of fishes in the ponds the farmer tries to provide maximum growing and development conditions for getting maximum number of fishes and thereby maximum number of fishes reach marketable sise.

Kinds of Operations

As already mentioned once the site is identified the farmer has to decide whether he wants to go for breeding and production of fry and fingerlings or rear them till they attain marketable sise. For breeding he needs more number of small ponds. The number of ponds may be more if he is trying to breed several breeds of fishes and in the same breed several sets of breeders (one set of breeders consists of two males and one female). If the farmer is planning to rear the fish only he needs big ponds. Depending on the number of fishes he wants to rear he may need one or more ponds. He also should consider whether he is raising the fish for the family usage or tor marketing purposes and whether he wants it as a main business or as a subsidiary business.

Questions like these have to be considered so that he can construct:

- The right kinds of ponds,
- The right number of ponds and
- Stock the right kinds of fishes.

Types of Ponds

The types of pond a farmer can build depend upon water supply, soil and topography. Generally, two types of ponds are built -barrage ponds and diversion ponds. The main difference between these two types of pond is the water source though in many aspects the construction details are the same.

Barrage Ponds

In barrage ponds the water source is the rain run-off water or a spring forming into a small stream running down the valley. The site is marked by a valley type of topography and the pond is constructed by building a wall across the valley and store the running water through the valley. The area where water is collected is called the pond area.

The barrage pond requires only one wall at the lower side of the pond area. By building a number of walls at suitable intervals the farmer can make several ponds successively one after the other forming a chain of ponds. The water enters first into the pond which is at the highest elevation among all the ponds and then passes on to the next pond lower in elevation. In other words the drained out water from the higher pond is collected by the lower one through a drainage pipe and then to the next lower till we reach the last one. From the last one the water is drained out safely. When ponds and the water inlet and outlet are arranged in such a way that the water flows from one pond to the other it is called ponds in series or rosary system. Water can also be directed independently into each pond from a channel running along one side of the pond and drained out into another channel at the other side of the pond. This is the best system of water supply to the fish ponds. Arrangement of ponds in this way is called ponds in parallel system. The channel that supplies water is called inlet channel and the channel that collects water from each pond is called drainage channel.

The drainage outlet may be fixed at the bottom side of the pond if the pond is to be drained out completely. Besides this drainage outlet there should bean outlet at a height beyond which the farmer does not want to accumulate the water in the pond. This is for the protection of the build which will be broken if water gets accumulated beyond certain height. The excess water is drained out from the pond through this drainage which acts as an overflow channel. The drainage and the overflow channel should be constructed in such a way that the water should be drained out safely without eroding the soil and causing damage to the pond.

In barrage ponds, flow channels are fixed at a suitable height in order to drain out the excess water that may be accumulated during the rain. There are many variations of overflow channel system from which the farmer can choose the most suitable one. The drainage and overflow channel can be joined together just before they join the common big drainage channel.

Barrage ponds should not be built where the flow of the water is high. The walls will not be able to withstand the force of rushing water especially during the rainy season.

Diversion Ponds

Diversion ponds are those which receive water from river or stream through a diversion channel. One diversion channel takes water to the pond and another diversion channel act as a drainage. The drainage channel may be taking the water into the same river or stream from which the water came in or to a farm where it is used for irrigation or the water may be drained out into any other outlet. If more than one pond are constructed they are connected with the inlets and outlets in series or in parallel design.

Diversion pond can be made in a number of ways depending on the ingenuity of the farmer. It can be built anywhere provided water can be taken to it from a water source by channel or pipe.

If the pond is built on a flat ground there should be walls all around to prevent the collapse of the sides. Sometimes the ponds are constructed by deepening an area which is lower than the surroundings. If there is enough space and water supply is assured there can be several ponds successively constructed to use the overflow of water from one diversion pond to a second one and from the second to the third and so on, till the last.

Whatever may be the type of pond being constructed there will always be some advantages and disadvantages. The farmer has to consider both the advantages and disadvantages and make his decision most beneficial to him. However, diversion ponds are better than barrage ponds because they are less likely to overflow and the water source is more dependable. But barrage ponds are cheaper in construction.

Similarly parallel system of pond construction is better than the series or rosary system. One can have a mixture of barrage ponds and diversion ponds arranged in series (rosary) or parallel system. The art of planning and constructing fish ponds is very much an individual thing. There are many ways of using land and water resources. But the exact shape, number and size of the ponds have to be decided by the farmer.

Number of Ponds

The number of ponds usually depends on the number of functions the farmer wants to perform and also the number of times he would like to repeat a single function within a given period of time. The functions related to the fisheries are holding a breeding stock, breeding, hatching, raising fry, raising fingerlings and rearing for marketing. Normally these functions are better done in different ponds or water bodies.

At the same time it should be remembered that several functions can be performed in the same pond or water body. For example, breeding and hatching and raising fry can be done in the same pond. However, for breeding large number of breeding stock we require more than one pond. Similarly, the number of ponds also depends on the number of breeds or kinds of fish the farmer wants to breed. For example, if the farmer wants to breed several breeds of fish, different ponds may be required, or several sets of the same breed is to be maintained. In the same way if the frequency of breeding is more (eg. once a week, once in two weeks, a month etc.) the number of ponds required will be more. At least two ponds are required for rearing for market purposes. If the farmer cannot make more than one pond he can also breed and hatch the fish in an enclosure of inverted mosquito-net fixed inside the pond.

Size of the Pond

The size of the pond also depends on topography, water supply, and the need or volume of production one intends to have. However, the nursery ponds will be smaller than the rearing ponds. Apart from this the size of the rearing pond itself can be varying. The advantages of small ponds are, function like harvesting, draining, refilling, treatment for diseases, etc. can be done quickly and easily. But the cost of construction and the land area required for the small ponds will be more compared to large ponds. But for the small and marginal farmers few small ponds are better than having large ponds.

Depth of the Pond

The depth of the pond also depends on several factors such as breeding, fry raising, fingerling raising, rearing for marketing, etc. Least depth is required for breeding and fry raising and maximum depth is required for rearing for marketing. Normally for breeding and fry raising the pond depth ranges from 0. 5 to 1. 0 m while for raising fingerlings a depth of 1. 0 to 1. 5 m is recommended. For rearing fish to market size the recommended pond depth varies from 1. 2 to 2. 0 metre.

Normally, sunlight does not reach deeper than two metres and the growth of the planktons will be restricted. At the same time too shallow pond will heat up the water or too much vegetative growth may occur and consequently the available oxygen to the fish will be much less than the required especially during the night hours.

But in commercial farms breeding and raising of fry is done in plastic tubs, basins, glass jars or even in big mud pots. Glass vessels are better for this purpose because we can observe the fertilisation, hatching and the conditions of the fry well.

Single Pond Operation

If the farmer can have only one pond he could purchase or collect fry and fingerling and rear them to market sise. For this there should be regular supply of fry and fingerlings. He can also breed fish inside a netted area in the same pond. The farmer who is having single pond can go for culture of single breed or a mixture of several breeds. The former is called monoculture and the later polyculture. Both have advantages and disadvantages.

Monoculture

Monoculture is the fish culture of only one species. It is intensive in caring, feeding and other management practices. It is easier to perform all the management practices since only one type of fish is involved. In monoculture the farmer has greater control over the sise, age and sex. In mono culture fish can be harvested selectively by using nets of different mesh sise. In monoculture diseases and pests are easily controlled at the same time there is a possibility of complete elimination of the whole lot by some epidemic type of diseases and pests if they go out of control.

Polyculture

In polyculture different breeds of fish are mixed together in fixed proportion or at random. Polyculture is more closer to the natural method of growing fish and under this situation only the fittest ones will survive. Several small fishes may be eaten by the bigger ones or they could not compete with the bigger ones for food and space. This problem is more in the case of poly culture in which both carnivorous and herbivorous fishes are reared together.

At the same time a farmer can have a polyculture with proper combinations (in number) of selected breeds of fishes. The selection is done on the basis of feeding habits. Such polyculture is called composite fish farming. For example, a farmer can rear a mixture of Indian carps such as catla, rohu, mrigal and Chinese carps such as silver carp, common carp and grass carp. The catla and silver carp normally takes feed from the upper layers of the water body, the rohu and common carp from the middle layers, while the mrigal and grass carp takes feed from the bottom layers of the pond. All are herbivorous in their feeding habit. In the composite fish culture full utilisation of the pond is possible.

If the breed introduced are naturally fast multiplying like tilapia and their number is a problem in the pond, another fish breed like catfish which is carnivorous can be introduced. The catfish will feed on the organisms at the bottom as well as the fry of tilapia keeping the population of tilapia under control. Introducing some grass carp among other breeds of fish that are being reared is the best method of controlling the weed growth in the pond. For farmers who have only one pond and do not have facilities like regular supply of fry and fingerlings nor he is unable or unwilling to breed them and those who cannot feed and care in the scientific and intensive ways, it is advisable to go for poly-culture with fishes that naturally breed in the pond.

Monosex Culture

In monosex culture either male or female fish of the same breed or more than one breed is reared. The advantage is faster growth of the fish since all the energy is concentrated on the growth and development. No energy is used for the reproduction.

Sexes can be separated individually during the breeding season or by noting the difference between the male and female in a breed of fish. In the cross breeding of some fish the off springs produced are sterile by nature or produce hundred per cent male fish. For example, if we cross male tilapia of Tilapia macrochir with female of T. nilotica, the male of T. mossambica with the female of T. nilotica and male of T. hororum with female of T. mossambica the off springs will be completely male. There are no crosses in which 100 per cent off- springs produced are female.

Monosex culture is valuable but difficult for ordinary farmers to carry out the breeding and rearing. Therefore, it is not recommended generally to the small scale fish farms unless there is a steady supply of male fry or fingerlings.

More than One Pond Operation

If the farmer has more than one pond then, he can rear fry or fingerlings for marketing in all the ponds or he can take up the breeding in one, raise fry or fingerlings in another and can rear fish up to marketable size in another pond. If he still has more ponds he can replicate the same functions in more than one pond. With three ponds a farmer can go for breeding and raising of fry for his own farm. The smallest one is used for breeding while the others may be used for rearing till they attain marketable sise.

The major difference between a large farm operation and a small farm operation is the number of ponds and the fry or fingerlings reared. In big fish farms all the input requirements will be more so also the greater output can be expected in terms of production as well as income.

OTHER FACTORS

Besides the above mentioned factors other important considerations in planning for the fish farm are: marketing facilities, transport facilities, communication facilities, banking facilities, availability of fish feed at the local level, storage and processing facilities, ownership of the pond area, labour force for construction and maintenance of the fish pond and availability of equipments and implements.

Marketing Facilities

For commercial production availability of market facilities is an important consideration while planning for fish farming. Market facilities here means mainly the openings for sale of fish and the demand for it.

Market can be at the local level or at a distant place. The local market depends on whether people are fish eaters or not and if they are, whether they have the purchasing power or not. Even if the local people are fish eaters but they have no purchasing power there will not be any local market.

If the market is at a distant place the question is how far away, is it within the reach or far away? How do we take the fish to that place? Are there transport facilities and vehicles to transport the fish or not? If fishes are to be marketed how frequently one wants to or can market, daily, weekly, fort- nightly or monthly? For daily marketing the production level should be very high so that he can catch sizable quantity of fish for the market.

The quantity of the fish to be marketed is an important consideration in employing any transport system. Unless there is sufficient quantity of fish one cannot employ vehicles to transport the fish produced. Proper quantity of fish can be ensured if several fish farmers join together to do the marketing operations.

Marketing considerations also include the facilities for purchasing things required for the fish farmers not only things directly related to fish farming but also for their livelihood. If such marketing facilities are available greater will be the possibility of success.

Facilities for storage,and processing the fish is another consideration related to marketing. If there are such facilities the fish farmer need not worry about the sale of fish immediately after the production.

Transport Facilities

Whether already some transport facilities are existing or not is an important consideration for a farmer to plan for fish farming. Transport facilities include mainly motorable road, railways or waterway transport. If there are roads are they public or private? Are there vehicles plying'? Are these vehicles private or government? If regular commercial vehicles are not plying are these vehicles available on rent or can the fish farmer have a vehicle himself? For a successful fish farming business, proper road facilities are important.

Communication Facilities

At present there are many communication facilities available in the market. But at least telephones should be available to the fish farmer. Through telephones he can do lot of the business transactions and thus can save time, money and effort. Where written communication are to be transacted, at least post and telegraph facilities are needed if not fax machine facility.

Banking Facilities

Banking and credit facilities are essential for the fish farmer to begin his enterprise. Initial investments are very high for the establishment of the fish farm. The major expense is in the construction of the fish pond, the channels, and other related structures. Normally no one will have sufficient reserve of money to begin a fish pond construction. Therefore, he needs to take loans from the bank and such banking facilities should be available to the fish farmer.

Availability of Fish Feed

The most important among the running expenses for the fish farming is the expenses on feed. Feeding is an every day activity to achieve the targeted production of fish. The quality of feed in terms of protein percentage differs depending on whether it is a fry, fingerlings, rearing and fattening stage or breeder fish. The feed or the ingredients for the feed mixture should be available to the fish farmer at the local level.

If they are not available at the local level they should be available from the nearby market. This involves increase in the cost of production as he will have to pay for the transport. Whether at the 10callevel or at a distant market quality feed should be available to the fish farmer. Otherwise fish farming cannot be done and hence availability of feed is an important consideration in the planning for the fish farm.

Availability of Labour

For running a home level fish pond only the family labour is enough. But for commerciallevel fish production we need certain number of labourers to carry out the day-to-day operations such as feeding, checking of the water inlet and outlet, the quality of water, cleaning of the pond from excess aquatic plant growth, aerating the water with oxygen, pumping in or out of the water, etc. Labour are also needed for periodic operations like handling the breeders, breeding, hatching, raising fry and fingerlings, sale of try and fingerlings, harvesting of fish, grading and packing, transport of the fish, purchase or preparation of feed mixture etc. Therefore, one who plans for fish farming should consider the availability of the labourers and the cost of the labour.

Availability of Equipments

Fishing nets, double walled fish breeding net, hand net, vessels, tubs, aerators, feeders to provide feed for the fish, injection syringe and needles, mortar and pestle, saline solution bottle, dissection-box, etc. are the necessary equipments required for the fish farming. The fish farmer while planning should consider the possibility of easy access to these equipments either by purchase or by borrowing.

Storage and Processing

Often it is not possible to market the fish immediately after they are caught and there is a need for holding the stock for a day or two without the fish getting spoiled. Deep freezer cold storage and ice boxes are the necessary facilities for storing the harvested fish.

The harvested fish can also be stored alive in special enclosures or tanks with sufficient water and minimum feeding facilities if they are to be held for more than one or two days. These enclosures can be constructed like big box with low height opening by a lid fixed with hinges. They are best constructed with strong wire-mesh so that when they are placed in the pond with the harvested fish, the pond water itself will be available to the fish. Selected fish can be picked up from these enclosures as and when the customers come to purchase the fish. The customers should be able to see the fish and choose whichever ones he wants to buy. In such enclosures the farmer can catch a particular fish without disturbing the other fishes in the enclosure. The enclosures made of strong galvanised wire-mesh are fixed on to the frames and hung with the help of rope and pulley mechanism. The enclosure with the harvested fish is kept in the pond water. When the customer comes to purchase, the enclosure can be lifted sufficiently enough to see the fish inside and the customer can choose. To catch the fish the enclosure is lifted, the lid is opened and the particular fish is hand netted out and the enclosure is lowered again into the water after closing the lid. This is only one of the methods of storing live fish. Each fish farmer can device his own method depending on his resourcefulness.

Processing such as cleaning, dressing, icing, canning, pickling, converting into fish products and byproducts, etc. add value to the harvested fish. The value added this way may be as high as 100 per cent and this profit derived is in terms of net profit for the fish farmer as well as for the number of labourers who will be employed in the processing in terms of the job opportunities available to them. The fish waste generated during the processing can be utilised as one of the ingredients for the fish feed.

Ownership of the Land

The ownership of the land on which the fish pond is constructed is an important consideration. Several infrastructural facilities at the pond site cannot be installed if the land is belonging to a person other than the fish farmer who is only a tenant. The tenancy may be also for only one year or for few years after which he is not sure whether the tenancy will be renewed or not. The actual owner may not be interested in the development of the fishery. He may be satisfied with whatever he gets from the pond. A tenant fish farmer can only rear the fish till they attain marketable size and sell in the market. Under such conditions there is nothing much for the fish farmer to plan.

Whereas, if the land is belonging to the fish farmer himself he can have a long term plan to establish a fish farm with its full potential. Even if he does not execute the project immediately he can always have a full scale plan which can be completed stage by stage. Hence, under the full ownership of the land, planning has some meaning for the fish farmer.

Electricity and Power

In the modem world how important the availability of electrical power is, needs no mentioning. It plays a key role in every sphere of human life. The farmer, therefore, should check whether uninterrupted electricity is available or not at the pond area.

PREPARATION OF MASTER PLAN

Whatever so far discussed, are points related to the survey and assessment of the land and water resources and the facilities needed for establishing a successful fish farm. Once the survey is made a master plan in the form of a blue print is made showing all the details such as area of the total land coming under the pond system, number of ponds, size of each pond, shape of each pond, depth of the ponds, the water inlet channels, outlet channels, classiness to the house, road, railway and market facilities, the watchman's and the labourers houses, pump house, storage facilities, area for the preparation of the fish, ponds for breeding, hatching, raising try, fingerlings, rearing up to marketable size etc.

But in the case of a single or double pond at the home level use, such an elaborate blue print may not be required. However, a master plan roughly drawn on a paper will be of great help to get a clean idea of the ponds to be constructed and managed.

For the planning of commercial fish farms it is highly recommended to seek the help of a civil engineer and other technicians in the construction of the bunds, walls, silt catching structures, outlet and inlet construction with control valves,

Whether it is commercial enterprise or home level usage the ponds should be constructed according to the principles of the ecological and environmental maintenance. In short they should be environment friendly. The master plan also should include the cost-benefit ratio of the enterprise. This is especially true for the commercial enterprises. Even for the home level ponds also such cost-benefit calculation is very much encouraged.

Master plan in the case of commercial enterprises will show the stages of implementation so that such a plan can also be a guideline for the evaluation and monitoring of the project.

The master plan also should include the future management system, the system of monitoring and the type of records to be maintained on various activities.

Such a master plan with all the details is necessary to obtain financial assistance from the banks and other financing agencies. They are to be included along with the project to be submitted to the financing agencies.

The master plan also serve as a reference for the future for anyone who may be getting involved in the fish farming system.

ECOLOGICAL SUSTAINABILITY OF FISHERIES SYSTEM

Ecological sustainability incorporates firstly, the long standing concern for ensuring that harvests are sustainable, by avoiding depletion of the fish stocks. Secondly, ecological sustainability also incorporates the broader concern of maintaining the resource base and related specie at levels that do not foreclose future options. Socio-economic sustainability focuses on the macro level, that is, on maintaining or enhancing overall long-term socio-economic welfare. This socio-economic welfare is based on the blend of relevant economic and social indicators, focusing essentially on the generation of sustainable net benefits, a reasonable distribution of those benefits among the fishery participants, and maintenance of the system's overall viability within local and global economies. The socio-economic sustainability blends together economic criteria (such as the level of resources rent) and social criteria (such as overall distributional equity), recognising that these are inseparable at the policy level.

Community sustainability emphasises the group level, that is focusing on the desirability of sustaining communities as a valuable human system in their own right, more than simple collections of individuals. It is recognised that a community is more than a collection of individuals. Hence, emphasis is on maintaining or enhancing the group welfare of participating and affected communities including economic and socio-cultural welfare, overall cohesiveness, and the long-term health of the human system.

Finally, institutional sustainability involves maintaining suitable financial, administrative, and organisational capability in the long term, as a prerequisite for three components of sustainability. Institutional sustainability refers in particular to the sets of management rules and policy by which fisheries are governed. A key requirement in the pursuit of institutional sustainability is likely to be the manageability and enforceability of resource-use regulations.

As elaborated by Charles, the first three sustainability components can be viewed as the fundamental points of a Sustainable Triangle. The fourth, institutional sustainability, interacts among these potentially affected positively or negatively by any policy measure focused on ecological, socio-economic and/ or community sustainability. If each of the components is viewed as crucial to overall sustainability, it follows that sustainable development policy must serve to maintain reasonable levels of each. In the other words, system sustainability would decline through a policy seeking to increase one element at the expense of excessive reductions in any other. The concept of inseparable elements of sustainability in fisheries could be seen as the sustainability triangle.

IMPORTANCE OF SUSTAINABILITY INDICATORS

In this study, a formal methodology called Multi-Criteria Analysis (MCA) is used. According to Mendoza and Prabhu, MCA is a general approach that can be used to analyse complex problems involving multiple criteria, and also have advantages when applied in a complex and stochastic system like fisheries. At least there are three advantages of this method for fishery sustainability assessment.

First, it can deal with mixed sets of data, quantitative or qualitative, including stakeholders' opinions. Fisheries, as a system, are well known to be complex and stochastic so that incomplete information and understandings may exist. In this case, qualitative information from stakeholders, including experts groups, and experiential knowledge have distinct advantages for assessing sustainability indicators of fisheries system.

Secondly, the MCA approach also can be conveniently structured in order to enable a collaborative planning and decision-making environment. This environment provides an opportunity to develop such an accommodation for the involvement and participation of stakeholders in the sustainability assessment process. Finally, the MCA methodology is also still simple, intuitive, and transparent while it has strong technical and theoretical support in its procedure.

Analysis of Indicator Linkages

We have described the methodology of static and individual behaviour of each indicator from the stakeholders' perceived importance and conditions. However, it can be also argued that each indicator seldom affects the dynamics of fishery system and their sustainability individually. They are intricately linked and connected among them and consequently will affect the sustainability directly or indirectly. Their impacts are tied through a web of complex relationship which is difficult to extract on an individual indicator basis. Moreover, Prabhu *et al.* also mentioned that sustainability of a natural system can be compromised due to cross-criterion or cross-indicator interactions. In the other words, it would arguably be meaningful if analysis can be extended into behaviour assessment among indicators using system approach.

According to Mendoza and Prabhu, cross-indicator interactions can be analysed at different levels depending on the amount of information and knowledge about dynamic interactions between indicators. It can be analysed using quantitative system dynamics if sufficient information about each indicator can be identified. On the other hand, qualitative analysis and assessment of indicator linkages also can be more suitable in the case of lack of functional relationship between indicators. In this study, we used one qualitative method, that follows an approach with regard to problems of indicator linkages, called cognitive mapping. Cognitive mapping is included in the soft methodology category and differs from traditional formal methodologies in terms of the type of analysis and results generated. Generally, soft methodology results are descriptive rather than prescriptive. Cognitive mapping is a casually based mapping technique where concepts representing elements of a complex problem are organised and structured using arrow diagram. Arrows represent the connections and relationships among the indicators.

In this study, we defined two variables which Eden and Akermann considered as the essential variables of cognitive mapping method, *i. e.* , domain and centrality. Domain is an important factor of cognitive mapping because it reflects the density or the number of indicators directly linked to a particular indicator regardless of direction. That is, higher domain values of an indicator reflect a larger number of indicators directly affecting, or affected by, the indicator.

ANALYSIS OF SUSTAINABILITY INDICATORS

The first analysis for sustainability indicators is to generate a set of indicators in terms of their importance judged by a group of stakeholders. In this study, we used a set of sustainability indicators which is composed of four variable criteria of sustainability indicators, namely ecological-criterion indicators (5 indicators), economic-criterion indicators (5 indicators), community-criterion indicators (5 indicators) and policy-criterion indicators (3 indicators). A group of stakeholders that consists of three types of stakeholders was involved in this analysis, *i. e.* , fishers community (10 persons), fishery-related decision-makers (3 persons) and fish market-related stakeholders (2 persons). By using participatory assessment method, we discussed the indicators set and asked the stakeholders to define their value on indicator importance as well as their perceived target of indicators.

The first result is the acceptance rate of an indicator, that is the rate of stakeholder judgement to the indicator or whether the stakeholder involves the indicator in his/her judgement. From this result, we reveal that catch structure indicator (indicator a. 1) has the lowest acceptance rate, *i. e.* , out of 15 persons only 5 persons (or about 33 per cent) put their value to this indicator in their analysis. The participation rate for other indicators were calculated more than 33 per cent and maximum at the level of 100 per cent (all of stakeholders involve the indicator to their analysis). From the criterion level, the policy-criterion indicator group has the highest average acceptance rate, *i. e.* , all indicators in this criterion were involved by the stakeholders (100 per cent).

The following results show the importance of degree of indicators which are judged using a 7-point scale of values by the stakeholders. It is clear that according to stakeholder values, almost all the indicators are rated moderately to highly important. It can be seen from the average weight value, which is calculated in a range from 5. 15 to 6. 73, showing that almost of all the indicators are important. Furthermore, from the calculation of relative weights, we can clearly see that some indicators are rated lower than others.

For example, under the ecological sustainability indicators, indicator a. 5 (effect of human activities on the marine ecosystem) is rated lower than other indicators under the same variable (ecological sustainability). Similarly, indicator c. 11 (demographic indicator) and c. 12 (succession rate) under the community sustainability has low relative weight as compared with the other three indicators in the same variable. Moreover, we also reveal that among all of indicators, indicator c. 15 (security level) under the community indicators variable has the highest value of importance, while indicator b. 10 (fisheries employment/proportion of labour) under the economic sustainability variables has the lowest value of importance degree.

In overall, it is also clear that from the average total weight value, the stakeholders judged the policy and community indicator variables are more important than ecological and economic indicator variables. From this information we can say that rather than ecological and economic objectives, the stakeholders may argue that, policy and community objectives should be stressed more in the general fishery development policy. This is confirmed by the fact that fishery governance in Japan has emphasised on policy and community-based decision making for many years.

However, this finding does not mean that ecological and economic objectives are not important. A different picture is also found for the case of stakeholder group interests. The degree of importance in overall is different among group of fishers, policy makers and market-related persons. For example, in fishery group, community and policy sustainability indicator variables are important, while policy makers emphasise on ecology and policy sustainability variables, and market-related persons considered economic sustainability variables as more important variables. These comparative analyses can be used as guides in making decision as to what indicators are important and need to be monitored site-specifically.

From the calculation of standard deviation we also reveal that there is no complete agreement or consensus among stakeholders due to the standard deviation value of more than zero. In the other words, the opinions and judgements are more varied among the stakeholders. Higher standard deviation indicates diverging opinions; the larger the standard deviation, the more varied the opinions. Among fisheries sustainability indicators examined, ecological sustainability indicators has the highest average of standard deviation (1. 09), meaning that the judgement or opinions of the stakeholders for the ecological sustainability indicators are divergent and varied among others. On the other hand, due to its average standard deviation value (0. 75), it can be said that there is a degree of consensus among the stakeholders when making a judgement for the community sustainability indicators. The economic and policy sustainability indicators have average standard deviation value between those values. According to Mendoza and Prabhu, it would ideally be desirable to have some way of improving the consistency of the judgements (minimum standard deviation).

In addition to estimating the relative weights of indicators, the next part of analysis is to estimate the sustainable state elaborated from the perceived targets or conditions judged by the stakeholders. This analysis is also started by judgements of the stakeholders to score the perceived targets of each indicator followed by the calculation of Sustainability Index of Criteria (SIC).

We can see that sustainability index for the ecological indicators is the highest among other sustainability variables (SIC=3. 79). It is followed by the economy indicators (SIC=3. 57), community indicators (SIC=3. 26), and policy indicators (SIC=3. 20).

The results of SIC mean that based on their perceived value of condition for each indicator, the stakeholders judged that ecological indicator variable has a relatively best condition compared with other variables. This result can be complementarily discussed with the results of important judgement of indicators. Although, according to the stakeholders, policy and community indicator variables are the most important for fishery system in Yoron Island, they also judged that ecologically and economically, the fishery system in Yoron Island is relatively "sustainable".

THE SUSTAINABILITY OF SHRIMP FARMING

IS CURRENT SHRIMP FARMING PRACTICE SUSTAINABLE

The idea of sustainability has been interpreted widely to cover a wealth of different perspectives. Although there is much theory on the subject, very few authors have offered realistic and simple practical guidelines or methodologies for the assessment of sustainability. A general assessment of the sustainability of shrimp farming, based on this framework and the more detailed discussion in the chapter. Although it is not possible to go into all these these aspects here, some key elements will be considered.

It is often suggested that extensive or semi-intensive production is more sustainable than intensive production; indeed, it is commonly suggested that intensive production is unsustainable. This view derives from a narrow interpretation of sustainability, and confusion between correlation and causation. Sustainability depends not on the level of intensity, but rather on the quality of the site, the management, and the suitability of the technology to both site and management competence. Unfortunately, because of poor site selection and management, much shrimp farming in recent years has not been sustainable; and although some of the failures relating to intensive production have been spectacular, there have also been major problems with extensive and semi-intensive production. Different levels of intensity may be more or less sustainable in different ways; and different levels of intensity may be more or less sustainable in different economic or ecological contexts. Table compares intensive, semi-intensive, and extensive shrimp farming according to a range of different sustainability criteria related to some of the relevant areas.

The degree of sustainability assigned to each kind of shrimp farming will depend on the relative weight given to the different aspects, and the social, economic, resource, and ecological context in which development is taking place. It is impossible to generalise about sustainability of production, which will depend upon local circumstances (*e. g.* ,, seed availability; water source; feed supply; management; disease etc.). In the following sections different aspects of sustainability will be discussed in more detail, and will be related to intensity where relevant.

SUSTAINABILITY ASSESSMENT

Sustainability Assessment In assessing the sustainability of any enterprise or technology, consideration should be given to at least the following:

- The sustainability (or continuity) of supply, and quality of inputs;
- The social, environmental and economic costs of providing the inputs (eg depletion of resources elesewhere);
- The long term continuity (or sustainability) of production;
- Financial viability;
- Social impact;
- Environmental impact;
- The efficiency of conversion of resources into useful product.

In practice this amounts to a thorough economic (in the broadest sense) analysis, long term investment appraisal, and environmental impact assessment.

Table. Intensity and Some Aspects of Sustainability

Aspects of Sustainability	Extensive	Semi–Intensive	Intensive
financial viability (IRR, payback period)	low	medium	high
financial viability (profit margin)	low	high	medium
return on labour	low	medium	high
employment/ha	low	medium	high
land productivity (kg/ha; $/ha)	low	medium	high
water productivity (kg/m^3)	low	low–medium	low–high
nutrient conversion efficiency	high	medium	low–medium
energy efficiency	high	high	medium
effluent quality	high	medium	low–high
risk of production failure	medium	medium	high
long term sustainability of production	?	?	?

SUSTAINABILITY OF SUPPLY AND QUALITY OF INPUTS

Seed

In many countries seed is still caught in the wild. The supply of such seed cannot be guaranteed; its availability is generally seasonal; and it may carry disease. At high levels of exploitation, the collection of wild seed may have deleterious effects on wild capture fisheries (for shrimp and other species), and on the coastal ecosystem - either directly through removal of shrimp juveniles, or removal of other species in the by-catch. Hatchery produced seed on the other hand may also suffer from erratic supply related to disease or broodstock availability. The quality may be poor as a result of excessive use of chemicals, the practice of multiple spawnings, and poor feeding (eg related to a shortage or high cost of artemia cysts). In general however, hatchery production offers much greater potential for real long term sustainability and quality control.

Feed

While formula feeds are dependent on fishmeal there remains the question of the sustainability of exploitation of the source fishery, and the stability of supply. Imported feed degrades rapidly during storage in tropical countries and quality cannot be guaranteed. Formulation is still often more closely related to ingredient costs than nutritional and environmental requirements. In some cases, feed has been suspected as a source of disease. As noted above, the supply of Artemia cysts - mainly from the US and China - is erratic and quality variable.

Chemicals

The supply of chemicals is likely to remain relatively reliable and will probably increase. Quality is however frequently questionable. A huge range are now available to the shrimp farmer, especially in countries like Taiwan and Thailand, but neither their effectiveness nor their environmental impact are well documented. Heavy use of antibiotics may lead to disease resistance of pathogens of shrimp and possibly of humans. Heavy use of chlorine compounds for water or pond sterilisation may lead to the formation of toxic chloramines and other complex organics.

Skills

The skills required for aquaculture production increase steadily with increasing levels of intensity. Although it may be relatively easy to achieve very high rates of output on a virgin site, sustained production at high levels of intensity requires high skills levels.

SUSTAINABILITY OF OUTPUT

Disease

Disease is by far the most serious threat to the sustainability of shrimp farm production. In Asia the viruses MBV, Yellowhead, and whitespot have caused devastation in many countries, and a variety of Vibrio bacteria cause intermittent problems. It is possible that Thailand is currently experiencing a crash similar to that which devastated the Taiwanese industry in the late 80's and the Chinese industry in the late 90's. In Latin America Taura syndrome continues to cause serious mortality in many areas. Much has been said about the relationship between intensity and disease, but extensive and semi-intensive producers have been affected at least as much as intensive producers in recent years.

Feed Quality

Poor feed quality may reduce growth, depress health and survival, reduce water and pond soil quality, and increase pollution. It is also possible, though not proven, that it may transmit disease.

Water Quality

Poor water quality, caused by acid sulphate soil conditions, pollution (self, other farmers, other) or the presence of disease organisms may reduce growth, depress survival and introduce disease.

FINANCIAL SUSTAINABILITY

Feed Costs

Fish meal costs are likely to rise in the medium/long term as a result of steadily increasing demand, and steady or declining supply. Much will depend upon the rate of development of alternatives. Artemia used for hatchery production is likely to rise in cost as a result of steadily increased demand. At present there are very few major sources (US Great Salt Lake, China, and some production in Vietnam). Alternatives, including culture, are possible, but not competitive at the present time. Improved artificial substitutes are also possible but some natural food seems to be essential for good growth and survival.

Cost of Chemicals

The cost of chemicals is unlikely to rise substantially and new ones are constantly appearing on the market. Better site selection, design, husbandry, and water supply should reduce dependency on and cost of chemicals.

Labour Costs

Labour costs are related to the stage of development of the country. Less developed countries have a clear comparative advantage in this regard, but as they develop further this advantage will be reduced, and advantage in terms of other resources or sites will become more important.

Water Supply

High quality water supply may be very costly in several regards. A site with good water supply may be expensive or have other features which are undesirable and lead to extra production costs. The development of infrastructure for the supply of water is costly. Water pre-treatment is also costly. Furthermore, as the industry (and other industries) develops, the demand for good sites and high quality water will increase, and the supply may decrease (in terms of quality and quantity), leading to a further escalation of price.

Product Value and Markets

Product quality is increasingly important in all markets. Product quality may decline over time in poorly managed intensive systems as a result of disease, chemical residues, and smaller sized product. The average market price has been remarkably firm considering the steadily increased rates of production. However, there have been some unexpected oscillations.

At the present time there is a real danger of declining demand in American and European markets due to the poor poor environmental and social development image of shrimp farming. Many Americans now think that all shrimp farming involves the bulldosing of pristine mangrove, driving out local people, making a lot of money, using massive amounts of chemicals, spreading disease, and then abandoning a degraded environment. It is even possible that environmental groups will force a ban on the import of farmed shrimp from Asia unless it can be shown to be environmentally sustainable.

Following major growth in the 80s, prospects for market growth in Japan for black tiger shrimp (P. monodon) are limited. The US and European markets still have good growth potential, but only if the image of the product can be improved. The Asian market continues to grow well, and is less sensitive to environmental image. In terms of production effects on price, disease is likely to continue to cause problems in many areas, and act as a brake on excessive production and consequential reduced price. This is an interesting example of a sustainability conundrum: lack of sustainability among some producers may secure sustainability for others. Production from wild fisheries is likely to stay steady or decline.

SOCIO-ECONOMIC SUSTAINABILITY

A variety of socio-economic impacts have been described in different parts of the world, including displacement of local people as a result of increased land value; decreased land quality (eg salinisation; acidification; subsidence); and resource appropriation (government allocation through concessions of previously held common property resources, which had been used for a variety of subsistence uses by local people); and interference with navigational rights and conflict with the fishing industry. There is little doubt that these impacts have taken place in some areas, but by no means in all countries or areas.

It has also been suggested that the shift from subsistence agriculture and use of natural resources to employment on large fish farms may represent a decline in quality of life. This is a political, not a scientific issue and it is the job of scientists and economists to provide information on these issues rather than make value judgements.

ENVIRONMENTAL SUSTAINABILITY

Like all human activity shrimp farming has a variety of direct and indirect effects on the environment. Direct impacts include habitat destruction. Mangrove is the most frequently quoted example, but a wide variety of other coastal ecosystems may be lost or affected, including estuarine and saltmarsh habitat, and other kinds of wetland. Shrimp farming is also responsible for a variety of pollutants including organic matter (creating biological demand or BOD); nutrient enrichment (mainly nitrogen and phosphorus); and chemical discharges including antibiotics, pesticides/piscicides, and disinfectants.

A variety of indirect impacts can also be identified. The wild seed fishery may affect other species that depend upon this food source - including shrimp fishermen. It may also affect other species directly through the by-catch of species caught by default and discarded. Loss of habitat may also affect other fisheries such as small scale shrimp or mud crab fisheries, although extensive or semi-intensive ponds are probably a suitable habit for the latter species.

WHAT CAN BE DONE TO IMPROVE SUSTAINABILITY

Sustainability can be improved as a result of individual initiatives, cooperative initiatives, and government initiatives. If shrimp farming is to grow successfully and sustainably it is probable that all three types will be required, but government must play a pro-active role in facilitating such initiatives.

Site Selection and Farm Design

Site selection is up to individual entrepreneurs, but can be guided or encouraged by government through incentives, education and information. There are many documents offering guidance in this regard and the subject cannot be covered adequately here, but some general principles may be offered.

The landward fringe of mangrove or non-mangrove areas are generally better than pristine intertidal mangrove, both in terms of suitability for shrimp culture and in terms of lack of value for alternative activities or physical and ecological functions. However, it should be remembered that many coastal habitats other than mangrove have high value – indeed, many have higher value than much of the mangrove.

Many mangrove areas are of low suitability for aquaculture because of potential acid sulphate soils which become acidic upon exposure and drying, which normally takes place regularly in intensive and semi-intensive culture. In general clay or loam soils are most suitable (in terms of pH and water retention). Some fine sandy soils may also be appropriate.

The presence of other farms nearby is a strong negative factor because of the increased potential for disease spread or poor water quality. Well planned separation of water supply and disposal will ameliorate this problem, but disease may spread as wind borne aerosol and separation as far as possible is therefore desirable – if rarely possible. As a general principle there should be no, or limited, upstream or downstream production. Furthermore, the farmer must ensure that his/her own influent and effluent are clearly separated with effluent well downstream, or ideally entering a completely separate water system. The provision of appropriate infrastructure should facilitate such design. Good access to supplies and markets is clearly desirable.

Inputs

Feed

Both the cost and the environmental impact of feed may be reduced through lower nitrogen and phosphorus content. Lower protein content is possible if the quality is good and the content of essential amino acids such as methionine and lysine is adequate. Fish meal substitutes should be developed for these reasons, and also to reduce pressure on the capture fisheries which supply the product.

Extensive systems have the advantage of maximizing the use of natural food. There is some evidence however that highly intensive closed systems may offer the shrimp a wide variety of natural organisms, including plankton and bacteria, and that this may represent a further opportunity for reducing the protein content of feeds.

Chemicals

Better husbandry, better quality water, better quality feed, and better quality seed will all reduce the likelihood of disease and the need for and use of chemicals.

Seed

An increase in hatchery produced seed will reduce impact on wild stocks and allow a greater degree of quality control. Improved hatchery practice in terms of avoiding multiple spawnings, and better larval feed hatchery should be encouraged, perhaps through quality testing of seed. The introduction of disease certification is also technically possible and should be increased. Closing the breeding cycle rather than using wild spawners will allow for genetic improvement in terms of growth, food conversion aand disease resistance.

Producing seed of alternative species – in particular P indicus and P merguensis – should also be encouraged. Different species show different levels of resistance to most of the common diseases, and production of a greater variety – either on different farms, or as part of alternating cycles on the same farm – will reduce risk.

HUSBANDRY AND MANAGEMENT

Good management and husbandry can have a major effect on all aspects of the sustainability of shrimp farming.

Pond and Water Management

In Thailand in recent years there has been a significant shift from high exchange (up to 20 per cent of the pond water was exchanged each day) to more closed systems. There are three main types: semi-closed, where water exchange is minimized at the beginning of the production cycle and only changed as required towards the end of the production cycle; closed recycle, where production pond water is discharged to a settling pond/reservoir before being recycled into the production pond; and closed pond, where minimal water exchange takes place in the production pond, and water quality is maintained using intensive aeration and manipulation with a variety of chemicals and additives.

The management of pond soil and sediments in intensive systems is also crucial to sustainability in terms of both water quality and environmental impact. Harvesting of production using a seine net may stir up pond bottom deposits rich in organic matter and lead to the release of a pulse of highly concentrated efflluent to the environment. Harvesting with a bag net at the outlet while draining the pond will on the other hand minimize disturbance.

Similarly the removal of pond sediments using high pressure hoses will create a pulse of pollution, while drying and removal will minimize environmental impact, and may lead to the production of usable fertilizer or soil conditioner.

FINANCIAL VIABILITY

Reduce Costs

The main production costs in shrimp culture are feed, seed, and labour, with feed costs sometimes amounting to 50 per cent of total operating costs. The greatest potential for the reduction of costs therefore lies in improved food conversion ratio. This can be achieved through good husbandry and possibly through improved feeding technology. Lin (1995) showed that smaller scale farmers in Thailand achived consistently lower FCR than larger farms, presumably through better feeding practice.

In Taiwan some farms now use feeding trays to supply all food, and reduce feeding losses to the sediment. Staff training and extension should contribute to better husbandry and reduced feed costs.

Reduction in seed costs may depend on a competitive hatchery industry. Thailand's highly decentralised private hatchery network, with specialists for different stages of production ensures a highly competitive industry with relatively cheap seed.

Reduction in labour costs is not necessarily desirable. Indeed, a local economic objective may be to minimize all costs other than labour. Cooperative purchase may also help in bringing down input costs

Increase Farm Gate Value

There are many different components to farm-gate value: the quality of the product, the image and promotion of the product in international markets, distribution and processing costs and mark-up, supply and demand trends in international markets. Shrimp is an international product, and promotion should take place at an international level.

Producers worldwide should therefore collaborate through their representative bodies (eg the Asian Shrimp Council) to promote shrimp, and initiate quality labelling initiatives and guarantees to consumers of environmentally sound production practices. National infrastructure should reduce distribution and processing costs, but these are in any case generally scale dependent, and as the industry grows thay are likely to decline. Cooperative ventures on the part of producers may also facilitate marketing.

DIVERSIFICATION

There is enormous potential for diversification into other species – either as an alternative to shrimp, and P monodon in particular, as part of an alternating cropping cycle including shrimp, or in conjunction with shrimp in polyculture systems. All of these approaches are likely to reduce the impact and severity of disease outbreaks and water quality problems. Candidates include edible and pearl oysters, seabass, grouper, P. indicus, merguensis and japonicus, and a variety of seaweeds.

PLANNING AND DEVELOPMENT

There is little prospect of sustainable shrimp farming without adequate and implementable planning for the industry. The disasters which have afflicted many areas in many countries can almost all be blamed on erratic and unplanned development with inadequate or poorly designed water supply, and in particular on mixing of influent and effluent water on individual farms and between farms. In Thailand the average farm shares its water supply with 30 others (ADB/NACA 1995).

Planning must include:

- Identification of the best areas for shrimp and other aquaculture;
- Identification of value for other functions (eg nursery, protection, biodiversity, industrial development);
- Identification and discussion with stakeholders;
- Allocation/definition of development zones (aquaculture; other)

These zones must then be provided with adequate:

- Infrastructure in the form of water supply and disposal ensuring separation of farm influent and effluent, and possbly including some form of centralised water treatment;
- Extension, R&D;
- Marketing/processing/product quality labelling and certification initiatives;
- Hatchery and export product quality and disease certification schemes

However, such planning and support will have little effect unless farmers and entrepreneurs can be persuaded to operate within these zones and according to the rules. In this regard, incentives are generally more effective than constraints. A wide variety of measures may be used, including differential grants, credit, taxes or tax breaks, licenses, and concessions. These can be designed specifically to reinforce planning and zoning. In addition some regulations must be required, particularly those related to the control or regulation of stock movements and introductions.

ROLE OF GOVERNMENT, PRODUCERS, TRAINERS AND RESEARCHERS

Shrimp farming has the potential to be a highly profitable and sustainable industry, and the industry itself should therefore take on the burden of most of the above measures, at least in the longer term. However, the generally unsustained development of the industry in other countries demonstrates the need for a major leadership and facilitating role for government in the initial phase of development. This applies particularly to water supply and disposal infrastructure, which individual farmers are incapable of providing without the sophisticated organisation and large amounts of investment capital which are only available in a mature industry. Government on the other hand is in a position to organise and fund infrastructure development, and eventually charge for its use if appropriate.

Farmers themselves should however play a major role in many of the other initiatives required, in particular the labelling, disease certification, marketing initiatives. Academics, research and teaching institutions all have a potentially significant role to play in the development of sustainable shrimp farming; indeed the industry offers a large and expanding market for their products.

6

Methods of Fishing

Man has been using various methods to catch fish since the pre-historic times, and the fishing gear has undergone evolution in different parts of the world giving rise to various methods of the present day. The fishing gear along with the vessel, auxiliary equipment and men constitutes a "fishing unit". The size of a fishing unit is determined by the distance of fishing grounds from the shore, handling and disposal of catch as well as geographical factors. The entire period between the launching of a fishing gear and launching it again after a gap is called a "fishing cycle", and includes the period for launching, hauling, emptying the net, repair and getting ready for the next launch. The number of fishing cycles per day depends upon the daily pattern of occurrence (density) of fishes, types of fishing methods used, geographical conditions of the fishing grounds and the fishermen. It is difficult to enumerate all the fishing methods in use in various parts of India and there is no uniformity in the names given to various fishing gears used in commercial fisheries. Some of common methods are being enlisted here.

They are:

- Fishing without gear,
- Wounding gear,
- Stupefy- ing methods,
- Line fishing,
- Fishing by baited springs,
- Fish screens,
- Fish traps,
- Trap for jumping fish,
- Dip net or lift net,
- Cast net (ghagaria jal),
- Triangular net,
- Purse net,
- Drug net,
- Gill and drift nets, and
- Electrical fishing.

FISHING METHODS

Modern methods of catching fish all have analogues in traditional practices that have been used for thousands of years. As in modern fishing, the methods used historically were chosen because they were well suited to catching the targeted species. Spears and arrows, for example, were used when the target was a large fish that could be closely approached, either by stealth or by confining the fish in a trap of some sort. Poisons made from fruits or bark were used to kill many fish simultaneously if the fish were physically confined. Archaeological evidence makes it clear that many types of traps were used extensively in traditional fishing.

Active traps included baited box traps and hooks. Passive traps included weirs, which were typically fence-like structures made of a row of wooden stakes that had latticework woven in between them. They were positioned in estuaries, streams, and shallow rivers to intercept migratory fish (*e. g.* , salmon traps). Other passive traps included coastal fish ponds in which small fish were raised to a suitable size in a simple form of aquaculture (*e. g.* , Hawaiian fish ponds). Nets made from natural fibres date back to the Stone Age. Some were used in an active mode, *e. g.* , beach seines. Others were used passively, *e. g.* , gill nets. All of these traditional fishing methods are relatively unimportant in modern commercial fisheries. They are, however, still used in subsistence fishing and in near-shore and local fisheries.

MODERN ANALOGUES

The modern analogue of the spear and arrow is the harpoon. Actually the use of hand-thrown harpoons to catch whales dates back at least to the 12th century, when the Spanish were hunting Right Whales in the Bay of Biscay. Modern-day harpoons, fired from a cannon and armed with an explosive head, are used today by Japan, Greenland, Norway, and South Korea to kill roughly 1,000 Minke whales per year in so-called "scientific whaling".

The present International Whaling Commission moratorium on virtually all commercial whaling reflects the ability of canon-fired harpoons with explosive heads and a number of other technological developments in the whaling industry to decimate the once numerous stocks of large baleen and sperm whales.

Hooks and lines are still used to catch fish, but the infrastructure surrounding their use in commercial fishing is very different now than in the past. Hand lines are used in trolling, which involves dragging a baited hook through the water behind a boat. Fish caught by trolling include mackerel and some game fish.

Hand lines are also used in conjunction with live bait in a procedure called chumming. Live bait are thrown into the water to induce a feeding frenzy. The target fish strike at the hooks, which are sometimes baited and sometimes hidden in the feathers of artificial lures.

When the target fish are "hot" (*i. e.* , sufficiently excited by the presence of the chum) they will even strike at bare hooks. Examples of fish caught by chumming include skipjack and albacore tuna. The procedure is limited by the availability of live bait, which must be maintained in a viable condition in bait wells aboard the fishing vessel. Fresh seawater is constantly circulated through the bait wells. In the case of tuna, about one kilogram of chum is needed to catch 10 kilograms of tuna.

Trawl lines are another form of hook-and-line fishing. Trawl lines are very long lines (*i. e.* , a kilometer or more) to which hundreds of shorter lines with baited hooks are attached. Laid along the bottom, trawl lines target demersal fish and were once used exclusively to catch groundfish off the coast of New England. Cod, haddock, hake, halibut, perch, groupers and snappers are all targets of trawl line fishing. The technique is particularly useful where the bottom is too rugged for the use of nets.

Pelagic long lines are the analogue of trawl lines in the water column. The lines are supported by buoys and are set to depths ranging from roughly 50 to 250 meters. Pelagic long lines are much longer than trawl lines, with the main line being perhaps 60 to as much as 180 kilometers long. The main line is actually composed of 400 to 450 sections, with each section having a length of 150 to 400 meters. Typically five branch lines with hooks form one section. As many as 2,000 hooks can be operated in one set. Paying out the line from the stern of the boat requires about four hours and usually starts before sunrise. Hauling in the line begins in the early afternoon and is facilitated with the use of a line hauler. The retrieval process can take 10 hours or more. Catch rates are typically 3-4 fish per 100 hooks. The crew's work is not finished until the catch has been prepared and sorted, and a work day can easily last 18 hours. The long work day combined with the fact that pelagic long liners stay at sea more than 200 days per year has made finding crews difficult. Yellowfin tuna account for about half and albacore tuna about one-third of the fish caught with pelagic long lines. Other fish caught by this method include bigeye and bluefin tuna, sharks, marlin, swordfish, and sailfish. In the case of swordfish, which feed at night, the lines are set in the afternoon and picked up the next morning.

Traps account for very little of the present day world fish catch, but they are important in certain fisheries. Lobster pots and crab pots, for example, are routinely used to catch lobsters and crabs, respectively. They are basically an enclosed framework of wire that can be easily entered by the target animal but from which escape is prevented by a retarding device. The bait consists of meat scraps such as fish parts or chicken necks. Placed on the bottom, the pots must be pulled and checked on a regular basis since the target animal, once trapped, has nothing to eat. Alternatively, if more than one crab or lobster enters the pot, the animals may become cannibalistic.

A variation on more traditional traps is the fish aggregation device or FAD. FADs are basically floating objects and typically amount to little more than a simple float moored to the sea floor, in some cases in water several thousand meters deep. Many fish are attracted to floating objects, and it is not uncommon to find numerous fish congregating beneath or near them. Why fish are attracted to FADs is a topic of debate, and the answer may be that not all species are attracted to FADs for the same reason. Tuna, for example, tend to ignore FADs unless prey are present. Other fish, such as mahimahi, seem to be attracted to the FAD itself. By mooring FADs at well defined locations and perhaps equipping them with radio transmitters, governmental agencies can greatly facilitate the job of finding pelagic fish. Mahimahi, tuna, and billfish are among the fish whose catch has been facilitated with FADs.

Weirs and pound nets continue to be used to capture finfish along the shoreline. Pound nets, for example, are used in the Mediterranean to catch tuna, along both coasts of the North Pacific for salmon, and in the western Baltic Sea to capture eels and herring. The word pound reflects the idea that the nets impound the fish, which are easily removed from the confinement of the crib. Interesting variations on the more traditional traps are the aerial traps used to catch flying fish. Once the fish have been stirred up, they are caught in the air with special gear called veranda nets. This technique is used, for example, by South Pacific islanders, who catch flying fish at night by attracting them with torches.

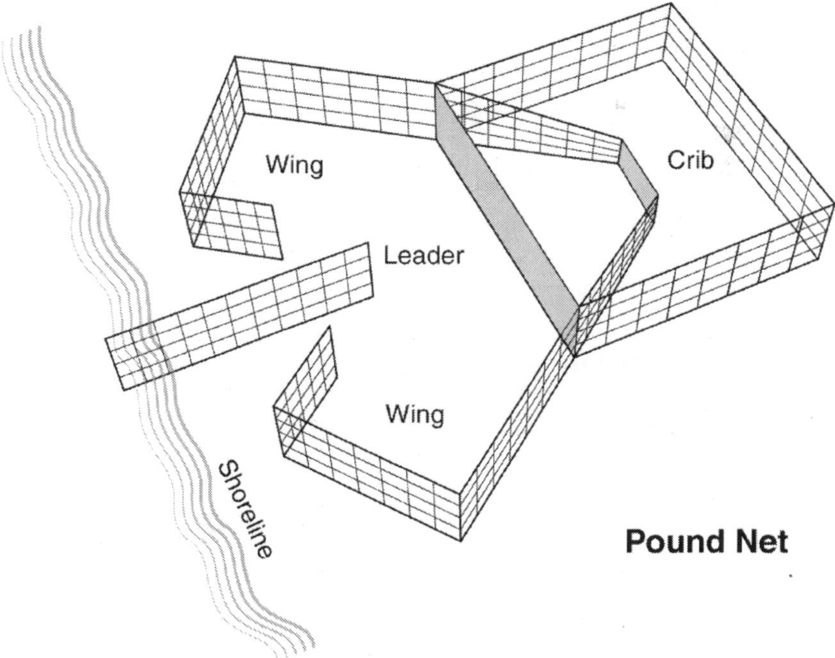

Fig.: Typical configuration of a pound net used to intercept fish migrating along a shoreline.

Nets of one sort or another account for most of the fish caught in modern commercial fishing. Gill nets are passive devices that either capture the fish in their meshes or entangle them. When anchored into the bottom they may be used to catch ground fish. More commonly they are suspended vertically in the water column by a series of floats and weights. In the latter mode they are referred to as drift nets.

The practice of using drift nets to catch fish has become highly controversial because the nets are simply too efficient. Unfortunately they often catch many fish and other organisms (*e. g.* , birds, marine mammals) that are not the target of the fishery. In the North Pacific, Japan, Taiwan, and South Korea for many years were engaged in a high seas drift net fishery that targeted squid, salmon, and billfish.

Every night during the fishing season, a single vessel would set a 10-30 kilometer long drift net that was picked up the following morning. Eisenbud has estimated that the length of all the nets used in the fishery totaled 33,000 km, roughly 80% of the circumference of the Earth. The use of high seas drift nets proved so destructive to both target and non-target species that in 1992 the United Nations General Assembly adopted a resolution requiring all nations involved in high seas drift net fishing to cease such operations by December 31 of that year. Nevertheless, drift nets continue to be used in many coastal fisheries.

The use of trawl nets (not to be confused with trawl lines) is an active form of fishing that involves pulling a net through the water or along the bottom to capture fish. An obviously limiting factor in the use of this technique is the location of the fish. If the target fish are demersal, then it makes sense to pull the net along the bottom.

However, dragging a net across the bottom is a difficult job, and the use of trawl nets to catch bottom fish did not become popular until steam and later fossil fuel powered vessels became standard fishing technology. Damage to nets was also a serious problem as long as the nets were made of natural fibres. Nets made of synthetic fibres proved far more durable for bottom trawling, but the technology is still impractical if the bottom is rugged or rocky. Even under ideal conditions, one of the early problems with the use of trawl nets was keeping the mouth of the net open as it was pulled through the water. This problem was initially solved by placing a beam across the mouth of the net, but this solution imposed a limit on the size of the net that could be used, since it is difficult to manage a large beam.

This problem was resolved by the development of the otter trawl, which was first used by the British during the latter half of the 19th century and later by Americans during the first half of the 20th century. The otter[1] trawl employs two broad, flat panels that are attached by lead lines called sweeps to opposite sides of the mouth of the trawl net.

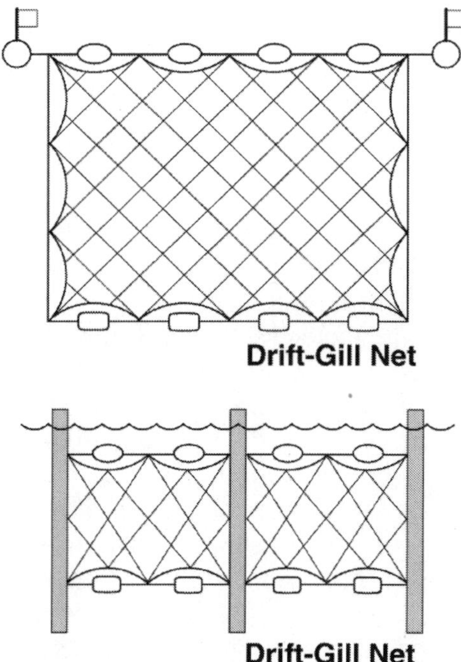

Fig.: Illustration of drift gill net (upper panel) and stake gill net (lower panel).

The net is towed by lines called warps attached to these panels, and the force of water against the panels as the net is pulled through the water pushes the panels apart and keeps the mouth of the net open. This ingenious strategy facilitated the use of much larger nets than would have been possible with the beam trawl design. The mouth of a trawl net can range form 10 meters to as much as 50-120 meters across.

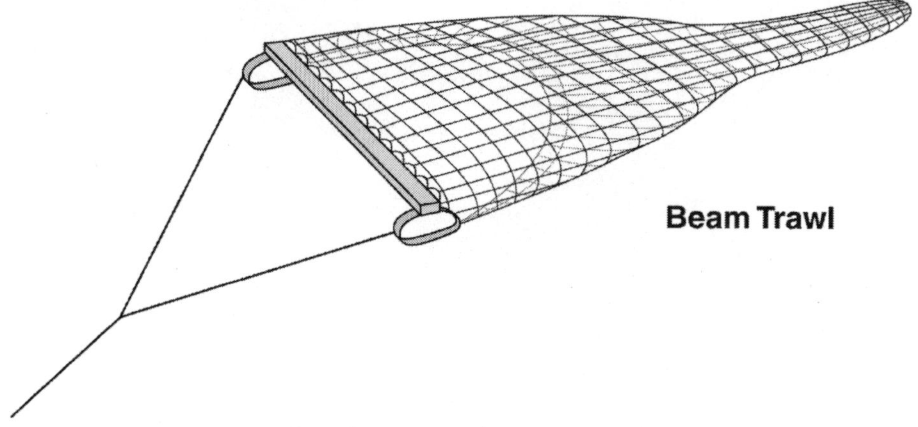

Fig.: Illustration of a beam trawl.

Otter trawls are used to catch both bottom fish and pelagic fish. In the case of bottom fish, the targets include shrimp, cod, haddock, whiting, flounder, saithe, and fluke. Limiting factors are the depth and topography of the bottom. It is not very practical to go bottom trawling in a water column 5 km deep or if the bottom is covered with topographical features that do serious damage to the net.

In recent years an additional concern has been the impact of repeated bottom trawling on the benthic habitat. It has been argued, not illogically, that repeatedly dragging heavy nets over the seafloor does serious damage to the benthic habitat, not unlike the effects of clear cutting a forest. In the case of mid-water fish, an important limiting factor is the ability of the fisherman to know at what depth the net should be towed. Prior to the development of the echo-sounder, a mid-water trawl was essentially fishing in the dark. However, with the advent of this device, it became possible to locate schools of mid-water fish from a surface vessel, and by placing a depth recording device on the otter boards, the fisherman could control the depth of the tow and dramatically improve his chances of catching fish. Modern trawl nets are even more sophisticated. They are equipped with sonar arrays that track the school of fish just ahead of the net. Bow sonar and computers automatically steer the ship towards the largest schools of fish. Fish targeted by mid-water trawls include cod, haddock, and hake, which sometimes aggregate in large schools far enough above the bottom to avoid capture by conventional bottom trawls. Trawl nets are now second only to purse seines with respect to the quantity of fish caught each year on a global basis, and in some parts of the world (*e. g.* , northwestern Europe) they account for more of the commercial fish catch than any other fishing method.

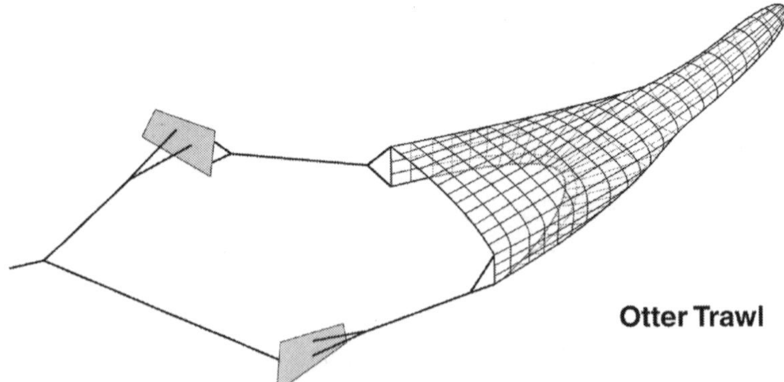

Fig.: Illustration of an otter trawl.

Globally by far the most important technique for catching fish is the purse seine. Purse seines are used to catch fish that school near the surface. The device is a large net as much as 1. 0 km long that extends perhaps 100 meters down into the water column. The strategy is to first encircle a school of fish with the net. The bottom of the net is then closed by pursing the net, *i. e.* , by drawing in a rope that passes through a series of rings attached to the bottom of the net. This effectively seals off the bottom of the net, which is then pulled on board the boat until the fish are confined to a relatively small pocket or bunt from which they can be conveniently netted or (in the case of small fish) pumped into the hold of the fishing vessel.

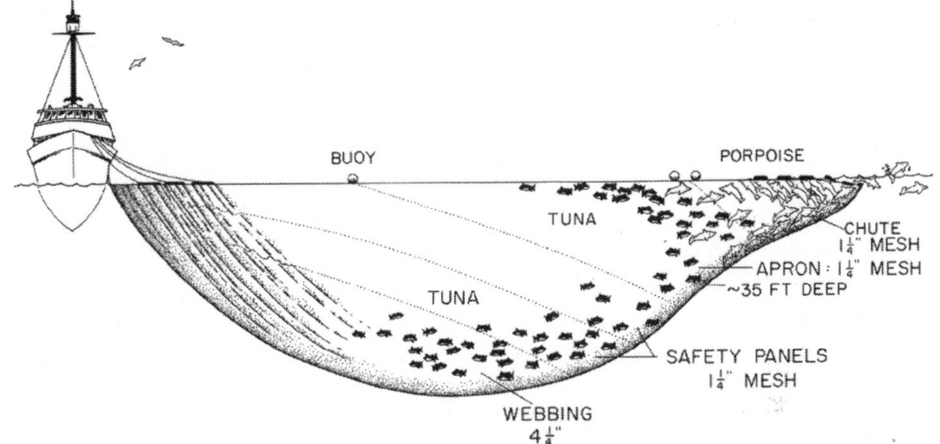

Fig.: Illustration of a purse seine.

Several factors limit the use of this technology. First, the target fish must aggregate in large schools at or near the surface. Second, there must be a practical means for deploying and retrieving the very large nets used in purse seining.

Prior to roughly 1950 it was common practice to set purse seine nets manually using small boats called dories. This technology (or lack thereof) limited the number of fish that could realistically be caught with a purse seine. This limitation was overcome with the use of the power block, which was invented by a California sardine and tuna fisherman named Mario Puretic. The device consists of a grooved pulley hung overboard from a boom and through which the entire net is drawn on board.

The pulley, or sheave, is driven by a compressed-air pump and has self-propelled rollers equipped with cleats that grip the net. Once the net passes through the pulley, it descends to the deck, where it neatly (more-or-less) folds down.

The third very practical problem is keeping the target fish in the net before it can be pursed. In the case of tuna, for example, the thermocline creates an effective barrier to escape.

As long as the net descends into the thermocline, the tuna are disinclined to dive to freedom. The shallow thermoclines of eastern ocean basins thus facilitate tuna fishing with purse seines. Some commercially important species caught with purse seines include anchovies, menhaden, sardines, pilchards, herring, mackerel, tuna, and salmon.

Comments and Observations

With this introduction to fishing methods, it is instructive to summarize their quantitative importance to global capture fisheries. At the present time roughly 50% of the catch is taken by purse seine. The principal targeted species include the Peruvian anchovy (Anchoveta), Chilean jack mackerel, Atlantic herring, Japanese anchovy, skipjack tuna, capelin, chub mackerel, yellowfin tuna, and European pilchard. The major nations and fishing areas involved in the capture of these fish. Most of these species are caught near coastlines by a small number of countries in a well defined geographical area. Exceptions include Atlantic herring, which are taken on both sides of the North Atlantic by five major fishing nations, and skipjack and yellowfin tuna, which are caught throughout the western Pacific and Indian Ocean (and elsewhere) by numerous nations. Of the purse seine catch, about 50% of fish such as herring, sardines, pilchards, and anchovies are used for reduction purposes. Tuna and mackerel go primarily for human consumption.

Otter trawling accounts for about 17% of the catch, the most important species being Alaska pollock, largehead hairtail, Atlantic cod, and blue whiting. The first three are used primarily for human consumption. The majority of the blue whiting catch is reduced to fishmeal.

The major nations and fishing areas involved in otter trawl fishing. The list includes two species, Alaska Pollock and Atlantic cod, whose catch has declined significantly in recent years. During the 1960's the Atlantic cod catch averaged about 3 Mt y^{-1}. In the last few years the catch has been about 1 Mt y^{-1}, primarily due to overfishing in the northwestern Atlantic. The catch of Alaska Pollock has likewise declined from a peak of 5-6 Mt y^{-1} during the 1970's and 1980's to about 3 Mt y^{-1} during the first few years of the 21st century. Of the remaining commercial catch, about 9% is taken with fishing lines, 8% with pound and other trap nets, and 6% with gill nets.

The choice of gear used in a particular fishery reflects a variety of considerations. Among these are the horizontal and vertical distributions of the target fish, their tendency to aggregate at certain times of year, and water clarity. Fish that congregate in shoals or schools near the surface are likely to be fished with purse seines. Tows or drift gear are used when the target fish are more dispersed.

Otter trawls or trawl lines are used to catch ground fish. In the case of shrimp, fishing (with otter trawls) is done at night, since the shrimp burrow in during the day. Otter trawls or pelagic long lines are used to catch midwater fish. In the case of the otter trawls, echo sounders are crucial to successful use of the gear. The behaviour of some fish warrants use of different types of gear during different seasons or even at different times of day. Herring in the North Sea, for example, are fished with bottom trawls during the day and with pelagic gear at night.

Both cod and herring form dense aggregations while spawning in the North Sea. Knowledge of the location of the spawning grounds and timing of the spawning event has historically facilitated capture of these fish. The migrations of salmon from the ocean to their freshwater spawning grounds are one of the best and historically well documented examples of the way fishing has been facilitated by human knowledge of fish behaviour.

Water clarity becomes a factor in fishing when the gear being used can be visually detected (and avoided) by the target fish. Gill nets, for example, are not very useful if they can be seen. It is for this reason that gill nets designed for use during daylight hours are made of synthetic material that is more-or-less invisible when hanging in the water. Historically, for example, purse seines were of little use for catching tuna around the Hawaiian Islands, because the water is clear and the thermocline deep. The clear water facilitated detection of the net by the fish, and the deep thermocline provided no barrier to escape before the net was pursed.

Finally, it is worth reiterating that technological developments have played a big part in the evolution of the commercial fishing industry. Echo sounders have been crucial to the location of schools of midwater fish. In the case of nets, synthetic fibres have dramatically changed the amount of time fishermen spend on the fishing grounds as opposed to mending their nets. Power, first in the form of steam and later in the form of fossil fuels, has had major impacts on the methodology and feasibility of fishing.

Pulling bottom trawls and hauling in purse seines is now done with relative ease using nets that would have been utterly impossible to manipulate by hand. An important technological development in the case of otter trawling was the use of stern ramps for retrieving the net after the trawl. As recently as the end of World War II otter trawls were still being "shot"[2] from the side of the fishing vessel.

At the end of the trawl, a boom was used to lift the cod end so that the catch could be discharged onto the deck or into the hold. The rationale for recovering the net from the side of the ship was concern over fouling the propeller if the net were brought on board from the stern. This problem was overcome by recessing the ship's screws. With this adjustment, it became possible to haul nets on board using a stern ramp, a development that permitted the use of much larger nets than had previously been possible. The idea of using stern ramps in this way undoubtedly derived from developments in the whaling industry, where stern ramps were routinely being used to haul aboard whale carcasses weighing as much as 100 tonnes.

The ability of boats to travel great distances to fishing grounds and to return safely with their catch directly reflects the availability of relatively cheap and reliable energy derived from fossil fuels to move the boats through the water. While long-distance fishing was by no means unheard of in the age of sailing, the prospects for fishing far from home are greatly enhanced when the fishing boat is powered with fossil fuel.

On such long fishing voyages, and indeed even when the ship returns to port after no more than a day or two at sea, preservation of the catch is a critical issue. Prior to the development of modern technology, standard methods for preserving the catch were drying or salting. In the former case the water content of the fish is reduced to a point where the multiplication of harmful bacteria is precluded. In the latter case the same effect is achieved by increasing the concentration of salt in the tissues of the fish. Historically fish were dried in the sun. The process works well with lean fish, but fatty fish such as herring, sardines, and anchovies are difficult to dry. The fat begins to go rancid before the water content is reduced sufficiently to block bacterial growth. The solution for fatty fish is salting, for which instructions can be found in cookbooks dating to the Roman Empire.

Drying and salting are still practiced in many parts of the world for cultural reasons and/or because more sophisticated methods of fish preservation are unavailable. Modern strategies for preserving fish involve either freezing or canning. The practice of canning is traceable to the early 19th century, when the French chef Nicolas Appert won a prize from his government for developing a new method of preserving foods that relied on heating under a seal. Initially the containers were made of glass. Preserving food in cans awaited the development of canning technology and advances in bacteriology that made canning a reliable method of food preservation. By 1900 canning fish had become a common practice, particularly with fatty fish such as salmon, sardines, anchovies, and later tuna.

Preserving fish by freezing initially relied on the use of ice, which became widely available for such purposes in countries such as the United States during the latter half of the 19th century. To maintain the quality of the product, the fish, once frozen, must be maintained in a frozen condition until it is thawed for cooking. A major advancement in freezing technology was the development of the plate freezer by Clarence Birdseye in 1929. With ice the temperature could at best be reduced to $0^{o}C$. With the freezing technology developed by Birdseye it was possible to deep freeze fish and thus ensure preservation from initial capture to the time the fish is thawed for cooking. The fish that freeze best are lean fish that do not can well such as cod and other so-called white fish.

This review of commercial fishing methods and technology has focused for the most part on large-scale company-owned commercial fishing. It is worthwhile to point out that substantial numbers of fish are still caught by small-scale artisanal fishermen using methods that are much less sophisticated and expensive than those employed by modern fishing boats. The provocative comparison of the two fishing industries.

Although the figure is based on an article published in 1980, the basic characteristics of the two fishing industries are accurately captured in this comparison. Far more persons are employed in the small-scale artisanal fishing industry, and per dollar invested in fishing boats, the small-scale industry is a much more effective way of keeping people employed. Virtually the entire catch from the small-scale industry goes for human consumption. Roughly 45% of the catch from the large-scale company-owned industry is used for reduction purposes.

For several decades there has been a real concern that the large-scale company-owned fishing industry would eventually eliminate most artianal fishermen. While the balance of competition favored the former during much of the 20th century, increases in the cost of fossil fuels may eventually shift the advantage to the small-scale industry. It is noteworthy that artisanal fishermen consume 5-10 times less fuel per tonne of fish caught than their counterparts in the large-scale company-owned industry. Many of the technological developments that company-owned fishing boats have been able to exploit to find and catch fish and bring them to market have made sense financially because of the low cost of fossil fuels. It is provocative to imagine how the commercial fishing industry would be structured if the price of oil were $100 per barrel, a scenario many persons now anticipate to occur sometime during the first half of the 21st century.

CONSTRUCTION AND MAINTENANCE OF A FISH FARM

A fish culturist needs different types of ponds for rearing various stages of fish, and has to decide on the layout of his farm and its fish and its extent. The design of the fish farm and the number and size of the ponds depends on the species of the fish to be cultured. The primary consideration in constructing a fish farm is the site which has to be selected on the basis of soil, water supply and drainage.

LOCATION OF A FISH FARM

Success in fish farming and economy of the construction would depend largely on the selection of a suitable site tor the farm.

The main considerations are:

- Topography,
- Soil type, and
- Water supply.

LAYOUT OF THE FISH FARM

Before starting the construction, the layout plans have to be drawn for the location design and the number of various types of ponds. A fish farm consists of 4 or 5 type of ponds, each for a specific purpose.

They are:

1. Hatching pits,
2. Nursing ponds,
3. Rearing ponds and
4. Stocking ponds.

Table. Arrangement for Different Types of Ponds in a 5 acre (2 ha) farm

Sl. No	Type of Pond	No. of Ponds	Size in Feet
1	Hatching pits	6	8′ x 4′ x 2′
2	Nursery ponds	4	50′ x 50′ x 4′
3	Rearing ponds	12	40′ x 30′ x 4′
4	Stocking ponds	6	300′ x 80′ x 6′

For a good fish pond, the slope of embankment should be 2:1.

POND MAINTENANCE AND IMPROVEMENT

The productivity of the pond depends upon its soil base, and can be greatly enhanced by controlling the vegetation, cleaning the pond bottom, lining, and fertilizer application.

ARTIFICIAL FEEDING

Fish production can be increased by artificial feeding. The fish feed should be simple and cheap. Whole grain flour, rice bran, oil cake, and kitchen waste are generally used as fish food.

Fish meal, meat meal, meat and blood is also use two. The food can be kept in baskets or spread on the water.

Formulation of artificial feeds for any fish requires a knowledge of nutritional requirement at different life stages ranging from larval to juvenile to grow out adult phases. Underfeeding can result in loss of production. Over feeding will cause a wastage of expensive feed and additionally a potential cause of water pollution, which is a major causative factor of loss of animals.

Fishing or Harvesting

This is done by draining the pond or netting. By draining, harvesting is complete and predators can be eliminated. Much less labour is required. The pond can be dried, cleaned, repaired. Soil can be enriched by fertilizer application. But all water is lost by draining. However if the ponds are constructed in a row, loss of water can be avoided and ponds are drained by turns.

Fishes attain table size, normally within one year of rearing period. Final harvesting is done by drag net either during summer months when the water level depletes to the danger level or after the monsoon, or when the market demands goes up. To have maximum catch of bottom dwellers from large and deep ponds, drag nets with pockets have been found to be more effective.

CASTING (FISHING)

In angling, casting is the act of throwing bait or a lure using a fishing line out over the water using a flexible fishing rod. The usual technique is for the angler to quickly flick the rod from behind towards the water. The term is also used when casting a net.

CASTING TECHNIQUES

Casting techniques vary with the type of fishing involved. Fly fishermen use artificial flies as a lure and use lighter rods and lines. They develop much finesse casting the flies, using motions of the hand and arm, so the flies land with great accuracy into or onto the water and mimic the behaviour of real flies. Salt water anglers usually use heavier rods and lines They often use lures and bait which are heavier than flies. Heavier again are the rods and lines used in surfcasting. Specialised, two-handed casting techniques are used to cast the lure or bait the added distances required in many cases to reach feeding inshore fish. In these casts the entire body, rather than just the arms, are utilised to deliver the cast, which may travel many hundreds of feet.

CASTING AS A SPORT

Casting is also a sport adjunct to fishing, much as shooting is to hunting. The sport is supervised by the International Casting Sport Federation (ICSF) which was founded in 1955 and currently has member associations in 29 countries. The ICSF sponsors tournaments and recognises world records for accuracy and distance. This sport uses plastic weights or hookless flies, and can be held on water or on athletic fields. There are competitive divisions for almost all types of fly, fixed spool and revolving spool tackle, and competitor classes. It is included in the World Games and has been considered for the Olympics.

The American Casting Association held its 100th Annual Casting Championships in 2008 at the Golden Gate Angling and Casting Club.

BOTTOM FISHING

Bottom fishing, called legering in the United Kingdom, is fishing the bottom of a body of water.

WHAT IS NEEDED

A common rig for fishing on the bottom is a weight tied to the end of the line, and a hook about an inch up line from the weight. The method can be used both with hand lines and rod fishing. The weight can also be used to cast or throw the line to an appropriate distance. Bottom fishing can be done both from boats and from the land. Bottom fishing targets groundfish such as sucker fish, bream, catfish, and crappie. Specialised fishing rods called "donkas" are also commonly used for bottom fishing. The objective for rigs used for bottom fishing is to take your bait to the bottom of the water and lure in the fish. The bait must appear appetising to the fish. The most common rig used in bottom fishing is called a "fish finder rig". The next rig is called a "porgy rig" for the reason that it is effective on porgies, grunts, snapper, and any other schooling, medium sized fish. A rig that is rarely used is called a "break-away rig". The final rig is called a "party boat rig" because you will see it on almost every party.

OTHER MEANING

In share market terminology, bottom fishing is buying the cheapest investments (in terms of valuation ratios) available. Bottom fishing is value investing concentrated on the very cheapest companies. The term can be derogatory as it can imply a lack of attention to the quality of the investments selected.

BLAST FISHING

Blast fishing or dynamite fishing is the practice of using explosives to stun or kill schools of fish for easy collection. This often illegal practice can be extremely destructive to the surrounding ecosystem, as the explosion often destroys the underlying habitat (such as coral reefs) that supports the fish. The frequently improvised nature of the explosives used also means danger for the fishermen as well, with accidents and injuries.

Although outlawed, the practice remains widespread in Southeast Asia, as well as in the Aegean Sea, and coastal Africa. In the Philippines, where the practice has been well-documented, blast fishing was known prior to World War I, as this activity is mentioned by Ernst Jünger in his book *Storm of Steel*. One 1999 report estimated that some 70,000 fishermen (12 per cent of the Philippines' total fishermen) engaged in the practice.

Extensive hard-to-patrol coastlines, the lure of lucrative, easy catches, and in some cases outright apathy or corruption on the part of local officials make enforcement of blast fishing bans an ongoing challenge for authorities.

Commercial dynamite or, more commonly, homemade bombs constructed using a glass bottle with layers of powdered potassium nitrate and pebbles or an ammonium nitrate and kerosene mixture are often employed. Such devices, though, may explode prematurely without warning, and have been known to injure or kill the person using them, or innocent bystanders.

Underwater shock waves produced by the explosion stun the fish and cause their swim bladders to rupture. This rupturing causes an abrupt loss of buoyancy; a small number of fish float to the surface, but most sink to the sea floor. The explosions indiscriminately kill large numbers of fish and other marine organisms in the vicinity and can damage or destroy the physical environment, including extensive damage to coral reefs.

IMPACT ON CORAL REEFS

Researchers believe that destructive fishing practices like blast fishing to be one of the biggest threat to the coral reef ecosystems. Blown up coral reefs are no more than rubble fields. The long-term impact associated with blast fishing is that there is no natural recovery of the reefs. Coral reefs are less likely to recover from constant disturbance such as blast fishing than from small disturbance that does not change the physical environment. Blast fishing destroys the calcium carbonate coral skeletons and is one of the continual disruptions of coral reefs. In the Indo-Pacific, the practice of blast fishing is a main cause of coral reef degradation. As a result, weakened rubble fields are formed and fish habitat is reduced.

The damaged coral reefs from blast fishing lead to instant declines in fish species wealth and quantity. Explosives used in blast fishing not only kill fish but also destroy coral skeletons, creating unbalanced coral rubble. The elimination of the fish also eliminates the resilience of the coral reefs to climate change, further hindering their recovery. Single blasts cause reefs to recover over 5–10 years, while widespread blasting, as often practiced, transforms these biodiverse ecosystems into continuous unstable rubble.

STRATEGIES TO CONTROL BLAST FISHING

Community-based Enforcement

In Tanzania, one of the few methods to help manage blast fishing is a joint approach between fisheries officers and village committees. Working together, they help the enforcement agencies recognise offenders by patrolling the sea as well as providing information collected in the local villages. As a result, this has assisted the enforcement agencies to reduce the occurrence of fish blasting from an average of 8 per day to zero. It has also provided sustainable funding to continue the efficient patrols, a certified planning institution, and suitable training and information to prosecutors and judges.

Similar patrols employed in Indonesia and Philippines have reduced the amount of blast fishing occurrences there. Based on dialogue with stakeholder groups in Southeast Asia and people of Tanzania and Philippines, it is evident that firmer enforcement is an effective strategy in managing blast fishing. Many countries have laws regarding blast fishing, but they are not fully implemented. Effective management of Marine Protected Areas (MPAs) is key in the patrolling of illegal fishing areas.

Blast Detection System

This method involves a triangulation system of hydrophones one meter apart that is capable of detecting blast events and at the same time eliminating other sources of underwater noise. The goal of the system is to improve and assist the effectiveness of fisheries patrol. Based on tests performed in Malaysia from 7 to 15 July 2002, a total of 13 blasts were recorded with a directional uncertainty of 0. 2°. An electronic compass would limit the bearing uncertainty to 0. 2° while correcting for the local magnetic effects of ferrous metals, therefore making sure the precision of the system is high.

Similar triangulation systems of hydrophones can potentially locate single blast events within 30 m at a range of 10 km. The detector system can be mounted on a patrol boat to help locate a probable range of blasts. Two or more patrol boats would permit accurate triangulation of blast events. Such a method is also beneficial to enforcement agencies, as it offers stronger evidence to support convictions related to blast fishing.

COUNTRIES

Indonesia

Blast fishing in Indonesia has been around for over 50 years and continues to transform its one-of-a-kind coral reefs into desolate gray moonscapes, as fishermen continue to use explosives or cyanide to kill or stun their prey. Dive operators and conservationists say Indonesia is not doing enough to protect the waters off the Komodo Islands. They say enforcement declined following the exit of a U. S. ,-based conservation group that helped fight destructive fishing practices.

Coral Gardens that were among Asia's most spectacular dive sites, were the latest victim of bomb blasting despite being located inside the Komodo National Park, a 500,000-acre reserve and U. N. World Heritage Site. The use of bombs made with kerosene and fertilizer is very popular in the region. While previously Komodo was relatively protected by a cooperative undertaking with CI (Conservation International) since the Indonesian government has assumed responsibility for park protection, there has been an upsurge in bombing. During a recent visit to Crystal Boommie, it was found to be 60 per cent destroyed, with freshly overturned coral tables proving recent bombing.

In the market in the city of Makassar, an estimated 10 to 40 percent of the fish are caught in this manner. The local fishermen find the technique to be easier and more productive than traditional methods. The goal for the country has been to implement stricter polices and fisheries management programmes to limit the killing of the fish as well as the destruction to the marine ecosystem. Forty years ago, blast fishing was practiced with dynamite which was in plentiful supply after World War II. Today, fishermen mostly use homemade bombs that are made from bottles filled with an explosive mixture; weights are also added to make the bottle sink faster underwater. After the bomb explodes, the fish killed or stunned by the shock wave from the explosion are collected.

Philippines

A 1987 study concluded that blast fishing was then very widespread in the Philippines, estimating that 25 per cent of all municipal fish landings (equivalent to 250,000 metric tons per year) were from blast fishing. A study conducted in 2002 reported that destructive fishing methods had caused the degradation of about 70 per cent of Philippine coral reefs and reduced annual fisheries production by about 177,500 metric tons in the 1990s. In 2012, the director of the Philippine Bureau of Fisheries and Aquatic Resources declared an "all-out war" against dynamite fishing and other illegal fishing practices.

Tanzania

In northern Tanzania, blast fishing, which is illegal, has re-emerged in recent years as a key danger to its coral reefs. This has occurred even though major institutions like local communities and the district government have been put in place for enhanced fisheries management.

The damage of blast fishing in the area has contributed to unstable coral reefs, discouragement of tourism investors, and a threat to the habitat of coelacanths in the region. Other impacts of blast fishing in the area include reports that citizens have died or lost limbs due to the blasting. The northern part of the country has many beautiful beaches and uninhabited islands. However, many investors feel and tourists are discouraged due to the fish blasting.

In Tanzania, coral reefs are essential for both ecological and socio-economic reasons. They are full of fish, lobsters, prawns, crabs, octopuses, mollusks, and sea cucumbers. In addition, coral reefs are one of the major tourist attractions in Tanzania. The coastal tourism provides a living for the people as well as foreign currency for the country. However, there has been an increase in the people living along the coast which has led to a large demand for fisheries.

It has led to overexploitation and destructive fishing practices. Blast fishing has been practised in Tanzania since the 1960s. It was during the 1980s and 1990s that blast fishing was at its peak in Tanzania. For example, in Mnazi bay, Mtwara, 441 blasts were recorded in two months in 1996, and 100 blasts were witnessed through one 6 hour period in Mpovi reef.

FISHING GEARS AND METHODS

DEFINITIONS

Methods to catch fish and other aquatic resources, with or without a gear, have always been practiced. Although the fundamental principles, *i. e.* filtering the water, luring and outwitting the prey and hunting, are the basis for most of the fishing gears and methods used even today, gears and methods have changed significantly over time and their capture efficiency is obviously hardly comparable to that of prehistoric times. A fishing gear is the tool with which aquatic resources are captured, whereas the fishing method is how the gear is used. Gear also includes harvesting organisms when no particular gear (tool) is involved. Furthermore, the same fishing gear can be used in different ways. A common way to classify fishing gears and methods is based on the principles of how the fish or other prey are captured and, to a lesser extent, on the gear construction.

1. Surrounding nets (including purse seines)
2. Seine nets (including beach seines and Boat, Scottish/Danish seines)
3. Trawl nets (including Bottom: Beam, Otter and Pair trawls, and Midwater trawls: Otter and Pair trawls)
4. Dredges
5. Lift nets
6. Falling gears (including cast nets)
7. Gillnets and entangling nets (including set and drifting gillnets; trammel nets)
8. Traps (including pots, stow or bag nets, fixed traps)
9. Hooks and lines (including handlines, pole and lines, set or drifting longlines, trolling lines)
10. Grappling and wounding gears (including harpoons, spears, arrows, etc.)
11. Stupefying devices

This classification is being slightly modified to accommodate the most recent development of fishing gears and methods and will soon be published.

Fishing methods have continuously evolved throughout recorded history. Fishers are inventive and not afraid of trying new ideas. The opportunities for innovation have been especially good in recent decades with advances in fibre technology, mechanization of gear handling, improved performances of vessels and motorization, computer processing for gear design, navigation aids, fish detection to mention only a few technologies.

Whereas technological development of fishing gear and methods in the past was aimed to increase production, the present situation with many overfished stock, limited possibilities to expand fishing on underexploited resources and concerns about the environmental impact of fishing operation, gear development is now very much focussed on selective fishing and gears with less impact on the environment.

Artificial reefs are structures placed on the sea bottom to gather fish. They are either used alone or with Fish Aggregating Devices (FADs). Most artificial reefs are large, permanent structures set in rather shallow water. Generally they are made of modern synthetics and hardware (concrete, metal, plastic pipes) which are sometimes finished using local vegetal materials (bamboo, coconut, leaves, coir for the anchor rope, etc.). In some cases lighter structures made from local materials are installed temporarily to lure fish to a specific area, during a certain fishing season.

Artificial reefs are established for various purposes:

- To enhance resources in coastal waters in order to facilitate exploitation;
- To create a biological reserve; and,
- To prevent the use of certain fishing gear, particularly bottom trawls, in a given area.

CONSTRUCTION AND MAINTENANCE

Different structures can be used as artificial reefs, such as wrecks, offshore oil rigs and pipelines on the sea bottom, heaps of oyster shells, etc. as well as specifically designed modules. Constructions covering a large area can be composed of individual modules, of which there are many models depending on the purpose.

For instance, to prevent trawling, a model might consist only of slabs of concrete with stakes. To enhance resources, that is to provide shelters for certain fishes and fish aggregation in general, modules should include holes (big enough to avoid being quickly sealed by marine organisms) where the material can facilitate the fixing of "fouling" organisms as food for aggregating fish. The overall dimensions of the reef and numbers and sizes of the holes for shelter are critical factors, as well as waves around it (modules should not bury themselves in bottom sediment). The modules should be bulky enough to project from the ocean bottom and massive enough to remain in place (even during severe storms); it might be suitable to anchor the modules of the artificial reef, particularly in shallow waters.

Material should resist rapid corrosion and should not introduce harmful substances into the marine environment.

Various elements can be used to create artificial reefs, including:

- Scrap material (car, old ships and barges, scrap concrete, household appliances, rock and rubble from excavations, etc.);
- Baled urban solid waste;
- Products such as sandbags and bamboo frames to construct temporary "reefs";
- Large concrete blocks, elements made from FRP and PVC; and,
- Offshore structures for oil or gas exploitation can also serve as artificial reefs and fish aggregating devices.

Building costs and installation factors (a very large number of artificial reefs must be set up) are also considerations in the choice of modules used. In shallower waters, setting FADs in conjunction with an artificial reef has proven profitable in several places. It has not been possible to identify a single model which could be universally recommended.

AGGREGATION PROCESS

An artificial reef plays the role of an offshore bank and protected area or park. The process of aggregation is often made in several steps: concretizing the blocks, then aggregating small fish and finally attracting larger fish.

The first few fish species arrive and eat organisms which have fixed on the artificial reefs which act as a food reservoir for demersal fish. Then, other species, such as predators of the former come.

For certain species (rock fish, octopus, crustaceans) artificial reefs are shelters; for small fish/juveniles, reefs provide protection from predators. Artificial reefs are also spawning areas for certain species (*e. g.* cephalopods) which explains how artificial reefs can enhance resources. In shallow waters shoals of small pelagic fish have been observed to use artificial reefs as stopping grounds.

A typical school of the sparid, *Montaxis grandoculis* (Mu), circling an artificial reef. This fish primarily eats molluscs.

IMPACT OF ARTIFICIAL REEFS

Enhancement and Recruitment

It takes weeks for a new artificial reef to attract fish and for the establishment of fish colonies/populations. Artificial reefs mainly concentrate on the remaining resource and the potential enhancement effect is, in general, considered quite low. In terms of habitat rehabilitation, artificial reefs have little, if any, success as they only concern a limited area.

It is worth observing that the assessment of some possible habitat rehabilitation and production enhancement would require a careful survey over a long period with regular data collection and careful monitoring of all fishing activities and in-depth knowledge of local ecosystems and ecological mechanisms. In most cases it is very difficult to evaluate fishing activity on artificial reefs. Therefore, regarding the potential socio-economic impact, there is a lack of thorough economic analyses of the costs and benefits of artificial reefs.

In general, the establishment of artificial reefs in coastal areas can facilitate smaller scale fishing activities with passive gears such as lines, pots and gillnets. In some places, recreational fishers also benefit from the creation of artificial reefs.

Biological

Regarding the biological impact of artificial reefs, two theories exist concerning the number of fish and fish species using reefs:

- Competition for limited resources, such as food and shelter is the chief factor; and,

- Fish mortality and recruitment are essential. The potential advantage from reefs could be a reduction in mortality (for instance by providing refuge from predators for small fish/juveniles) and/or recruitment stability.

Fish behaviour on artificial reefs differs from one species to another. This fact determines the different aggregating ability of the artificial reef for various species.

LEGAL ASPECTS AND REGULATION

Legal aspects related to artificial reefs include:

- Ownership of the artificial reef;
- Ownership of the fish on it;
- Limitation/prevention of certain fishing gear and therefore certain fisheries, resource utilization and management; and,
- Shipping and navigation.

A number of countries have set up strict regulations concerning the position of the artificial reefs, design and material used, proper marking, issue of a government authorization, amount of fishing gear authorized, regular reporting of fishing activities on artificial reefs, etc.

Any national regulation must take into consideration the provisions of UNCLOS (United Nations Convention on the Law of the Sea), conventions on navigational safety, shipping routes, conventions on dumping at sea, and decisions by fisheries management bodies.

When offshore structures for oil or gas exploitation act as a permanent fish aggregating device or artificial reef, they are covered by international standards and guidelines developed by the International Maritime Organization (IMO). The creation of artificial reefs has changed the perception of fishery resources from a primarily competitive hunting attitude to a collaborative rearing and nursing one.

In addition to direct impact of local fisheries, performances and revenues, artificial reefs have side-impacts on other fisheries and related activities in terms of catches and incomes. Where there is heavy or overexploitation, deployment of artificial reefs appears to become an alternative to more traditional development efforts of artisanal fisheries for effective near-shore fishery resources management. It is worth mentioning that the settlement of such structures introduces important changes in the management of fishing operations: *e. g.* fishing rights, working time allocation, relationship among fisher groups and between small and larger scale fisheries.

AQUACULTURE TECHNOLOGY

The technology employed in aquaculture has developed over many centuries but has done so more rapidly during the last half century. Aquaculture systems, and the technology used, vary from very simple systems, used for family ponds in tropical countries, where production is for domestic consumption, to high technology systems, such as intensive closed systems, like those used for rearing stripped bass. Herbivorous and filter feeding fish, reared mostly in simple systems of small freshwater ponds, however, account for about half of global aquaculture production.

Much of the technology used in aquaculture is relatively simple, amounting to small modifications that improve the growth and survival rates of the target species, such as providing additional food, adding seed animals collected elsewhere, managing water exchange to maintain adequate oxygen levels, and protecting the stock from predators. Greater understanding of complex interactions between nutrients, bacteria and cultured organisms, together with advances in hydrodynamics applied to pond and tank design, have enabled the development of closed systems. These have the advantage of isolating the aquaculture systems from natural aquatic systems, thus minimizing the risk of disease or genetic impacts on the external systems.

Developments in engineering, some learnt from offshore oilrig construction, increase the possibilities for offshore aquaculture using robust cages. Sea ranching, the release of young fish into the wild to improve the harvest in capture fisheries, has also made a start but its long term viability is still to be assessed. Major advances are also being made in the technology of the production of aquafeeds, which generally require the combining of a large number of ingredients into very small feed pellets.

Aquaculture Engineering

Aquaculture draws on well-established engineering fields for most of the design and construction needs of its production facilities. Building earth ponds is similar to building roads - a knowledge of the characteristics of soils and the limits of safe design are the basis of good construction. Similarly, the buildings used for hatcheries and other support activities are no different from those common in the housing, agricultural and commercial sectors. Sometimes ponds are lined with plastics or other impermeable materials, and here the techniques are similar to those for civil structures such as potable water reservoirs or sludge tanks.

The design and installation of water control gates, including in unstable soils, benefits from the long experience in this field in the agriculture and irrigation sectors. An understanding of hydrodynamics allows ponds and tanks to be built with good water circulation, oxygen mixing and without 'dead spots' where sediments might accumulate and cause health problems to fish.

For installations in the sea, the situation is somewhat different and many of the important engineering solutions, such as for fish cages or suspended shellfish growout systems, have had to be developed by aquaculturists themselves. They have benefited however from the accumulated knowledge of seafarers in general and fishermen in particular, in the design and operation of mooring and buoyage systems. More recently, when fish farmers have turned their attention to how to operate fish cages in locations further offshore where seas are rougher, the experience of the oil exploration industry has proven very valuable.

Techniques have been developed in recent years for the production of fish and other aquatic products in closed recirculation systems. To make these work, aquaculturists have needed to develop a knowledge of the biological processes operating - such as how bacteria can be used to neutralise and re-cycle the nitrogenous waste products produced by growing fish - and how to engineer the systems to meet the biological requirements. Knowledge of bio-engineering from the waste treatment and water treatment industries has made contributions to the development of closed aquaculture systems and there is probably more that could be usefully transferred from the sewage treatment sector to help solve problems in fish rearing.

Modern materials have brought many benefits to aquaculture. For instance, custom made plastic joints have simplified the construction of sea cages, and made them more reliable in high stress conditions. Experiments with huge free-floating or sunken net cages operated in the open ocean were begun several decades ago, for instance in the Caspian Sea. These showed some promise, but more reliable construction materials will make the farming of fish in such structures increasingly feasible. Modern materials and production methods have been important also, in the construction of plastic filter substrates for indoor recirculating systems. The fine detail of these has been found to make substantial differences to the efficiency of biological filters. In shrimp farms, specially designed matting materials that stand upright on pond bottoms, with a structure that promotes the growth of the small animals and plants that the shrimp can thrive on, have recently been developed and shown to boost production.

Engineering skills are important in the design of most aquaculture facilities and good engineering can affect the efficiency and economics of production. If capital costs can be minimised while still maximising productivity and reducing risk, the farming operation will be more profitable. Aquaculturists have proven very innovative over the past fifty years, constantly developing new technologies to support their farming operations. As new production methods and species for farming are developed, the engineering solutions needed to support them will continue to evolve.

Aquaculture Facilities

Aquaculture began with man making small modifications to natural habitats so as to improve the survival and growth of target species. Some of the oldest examples are in the rearing of freshwater fish in ponds, which has been practiced for thousands of years in Asia and at least for many centuries in Europe. The simple act of placing a mesh barrier across the outlet of a small pond or lake to prevent fish from escaping can make a big improvement in the supply of food. Similarly, along coasts, some tidal lagoons can be easily turned into ponds. Closing off such naturally occurring water bodies was the start, centuries ago, of much fish and shrimp aquaculture in Asia and, in modern times, in South America. Removing predators and improving the conditions within the pond, (for instance by providing more area of preferred water depth), supplying additional food and, later, by adding seed animals collected outside, were further steps that moved aquaculture production close to where it is today. The farming of seaweeds and molluscs (oysters, clams, mussels etc) developed similarly, as people made improvements such as providing more settlement areas for the young, or the removal of predators from the growing area.

Farmers had a natural desire to improve the productivity of their systems and, as knowledge grew, they learned to stock more animals, increase feeding and manage the exchange of water to maintain the conditions, such as adequate oxygen levels, that the animals needed to survive. The range of facilities used for aquaculture subsequently broadened for a number of reasons. Farmers encountered difficulties with more intensive rearing because of the uncontrolled influence of pond soils, local water quality, weather. Some of these problems could be resolved by rearing in ponds built of concrete, or lined with plastic, by bringing the ponds indoors under cover, or treating the water before flowing it to the culture ponds.

Secondly, as the naturally occurring ponds and lagoons all became used, prospective farmers had to take a broader approach and develop the technology and engineering to be able to use less naturally-favored sites. Net cages floating in protected coastal or inland waters were developed for fish culture, or fish were stocked in fenced areas of the sea or of large lakes. At the same time, the need for supplies of young animals (fry, seed) to stock these systems led to the development of hatchery techniques and dedicated hatchery facilities. For most species this aspect of production has proved more successful when conditions can be more closely controlled, for instance in indoor concrete and fibreglass tank systems, rather than in outdoor ponds. Often, the younger stages of aquatic animals are more sensitive than adults to physical and chemical conditions and these have to be managed within a smaller range, if production is to be successful. Thus hatchery facilities developed as a separate branch of the industry.

More recently, our knowledge has improved greatly regarding the complex interactions that occur in a rearing system, between nutrients, bacteria and the cultured organism. This and technological developments have allowed many aquatic organisms to be reared in completely closed recirculating facilities, including the farming of marine organisms at locations far from the sea. Closed systems have the added advantage of offering greater protection from the danger of disease entering from the natural environment and also of minimizing adverse effects of the production system on that external environment. Technology has also begun to open up the possibilities of growing fish in enclosures in the open ocean, something that could one day transform the nature of human food production on the planet. With 70% of the earth's surface covered by water, the potential is clear. Earlier technology has restricted the cage farming of fish to sheltered coastal waters. Here the number of available sites is limited, environmental damage is more likely and conflicts exist with other users. As cages are developed that can withstand the demanding conditions of the open ocean, or which can be operated below the ocean surface, farming could move further offshore.

A small start has been made, notably in Japan, through so-called 'ranching' programmes', the farming of the sea without enclosures where fry are released in large numbers into the ocean with the aim of improving returns from the capture fishery. Sometimes artificial reef structures are created underwater as well, to increase the available habitat and natural food for the fish.

Aquaculture Systems

Aquaculture - the growing of aquatic animals and plants - covers a wide range of species and methods. Some of the simplest production systems are the small family ponds in tropical countries where carp are reared for domestic consumption. At the other end of the scale are high technology systems, such as the intensive indoor closed units used in the USA for the rearing of striped bass or the sea cages used in Chile and Europe for growing salmon and bream. Nearly half of the world's aquaculture however is of herbivorous and filter-feeding fish such as carp.

Practised in freshwater, in ponds made of earth, and often managed by a single household, production systems are simple and in many cases have changed little over many centuries. A farmer will excavate a small pond near his house; sometimes, as for instance in the flood river deltas of countries like Vietnam, using the spoil to build a raised foundation for the house itself. The pond will typically be fed by natural rainfall or groundwater, or sometimes by diverting a nearby stream or irrigation canal. Some stocking occurs naturally, but more commonly a farmer will buy young fish from a breeder and stock them in his pond, often after holding them for a time in a small floating net known as a 'hapa' to check that they are in good condition before release in the pond. Production will improve if the farmer can stock a good balance of different species to make the best use of the varied kinds of food available in the pond. Typically the fish will be fed with household or agricultural by-products, and harvested when the family needs a meal.

Stocking and harvesting is often continuous, where knowledge of the best stocking level is learned from experience and where the pond is not drained for many years. For most such farmers, fish production is a secondary activity, a useful additional source of protein to add to the supply or income from his main agricultural or commercial activities.

Outdoor freshwater fish farming is practised commercially in many countries, including in the developed nations. The density of stocking may be much higher, the control of feeding, water quality and fish health more closely monitored, but even in intensively operated systems, the key parameters that need to be monitored and balanced for success are similar to those in the traditional household pond system.

Over the last half-century, systems for the indoor rearing of many species of fish have also been developed. Holding of fish in controlled conditions indoors has been important in the development of seed production (hatcheries). Also, as knowledge has increased of nutrient cycles, bacterial action and water chemistry, it has become possible to rear fish, both for food and for ornament, in 'closed' indoor systems, where the water is recirculated. Passing the water through filters that use bacteria to naturally break down and recycle the waste products produced by the fish allows a production system to be run in almost total isolation from the outside. Such systems have been important in the control of disease and also in maintaining the stable conditions that some species need to flourish and reproduce.

A fifth of current world aquaculture production is of plants - mainly seaweeds, that for the most part are grown on the seabed or on raft or racks in shallow coastal waters and used directly for food or for the production of alginate or carageenan (agar-agar). This sector of aquaculture is another largely run by small-scale growers, mainly in Asia but in some parts of the South America, who attach seedlings of marine plants to simple structures built close to shore and constructed from natural materials and then harvest the plants once they have grown. A further fifth of world production is of molluscs such as oysters, clams and mussels.

Again the majority of production is small-scale and by coastal people who often combine fishing activities with farming. Oysters and mussels are mainly grown on structures built above the seabed - poles or racks on the shore, or ropes suspended from rafts or floating lines. The farmer's role is to supply suitable places for seed to settle or to add to natural production by bringing seed from a hatchery - and then to maintain the conditions of waterflow and freedom from predators that the shellfish need to grow. Most commercially grown molluscs feed on microscopic algae floating in the water, so the farmer does not need to provide any feed.

Fish are also grown in coastal areas, in ponds or in floating cages. As technology advances, there is the potential to develop systems for rearing fish in the open ocean, either in sturdy cages or by so called 'ranching', where young fish are released to the wild and then collected by normal fishing or by training them to respond to specially generated sounds.

The tropical coastal zones of Asia and Latin America are where most shrimp farms are found. Shrimp farming is mainly carried out in earth-bottomed ponds built on flat land close to the sea. Shrimp are raised in ponds at a range of different densities. At low densities the systems need only limited inputs of feed and fertilizers. As the density increases, more feed has to be supplied and at the higher end of the range, so-called 'intensive' farming, machines have to be put in the ponds to mix and aerate the water.

Aquaculture is practised from the cold waters of the far north and south, where fish like salmon, arctic char and sturgeon are grown in ponds, flowing raceways and cages in the sea, down through the latitudes to the tropics, where carp and tilapia flourish in freshwater and shrimp and seabass are farmed along the coasts. It ranges from production of fish in naturally occurring ponds in rural areas to the intensive culture of ornamental fish in plastic tanks in the middle of a city. It is practised by the poorest farmers in developing countries as a livelihood and to supply much needed protein for their families - and by urban sports shop owners in Europe and the U. S. A. producing baitfish for weekend anglers. Systems can range from an intensive indoor system monitored with high-tech equipment through to the simple release of baby fish to the sea - but all with the same aim, of helping boost Nature's natural productivity

Biotechnology Development

Biotechnology - any technological application that uses biological systems, living organisms, or derivatives thereof, to make or modify products or processes for specific use 1994 UNEP Convention on Biological Diversity - has a wide range of useful applications in fisheries and aquaculture. It creates opportunities, for instance, to increase growth rate in farmed species, boost the nutritional value of aquafeeds, improve fish health, help restore and protect environments, extend the range of aquatic species and improve the management and conservation of wild stocks.

Some biotechnologies are simple with a long history of application such as the fertilization of ponds to increase feed availability. Others are more advanced and take advantage of increasing knowledge of molecular biology and genetics, *e. g.* genetic engineering and DNA disease diagnosis. The field of genetic biotechnology similarly ranges from simple techniques such as hybridization, to more complex processes such as the transfer of specific genes between species to create GMOs (genetically modified organisms).

Over the years, our knowledge of fish breeding requirements has improved and the ability to induce artificial breeding developed through the use of natural or synthetic hormones and/or environmental manipulations. (For example changing photoperiod or water temperature can induce some fish to spawn). These have been key factors facilitating the application of more advanced biotechnologies.

Selective breeding, the maintenance of stocks genetically improved by chromosome manipulation, line crossing, and sex reversal all depend on the controlled breeding of farmed species. These improvements in reproductive technologies have also assisted aquaculturists greatly in their efforts to domesticate aquatic animals. In addition, by making it possible to remove the natural constraints and timing of breeding, farmers are able to mate many more species at times that are most beneficial, and thus helping to ensure a steady and consistent supply of fish for consumption.

FEEDS TECHNOLOGY IN AQUACULTURE

Although aquaculture dates from the earliest parts of human history in Asia, Europe and in the Pacific Islands, it is only in the last few decades that the sector has begun to catch up with the rest of animal agriculture in terms of the science of feed milling and nutrition. Despite this, aquaculture currently represents the fastest growing segment of agriculture and the animal feed milling industry, particularly in China and the Asian region where over 90% of global aquaculture production is currently realized.

A very wide range of ingredients are used to prepare aquafeeds. Feeds range from single component feeds available on-farm such as grass or rice bran to farm-made formulated feeds and commercial feeds. They include aquatic and terrestrial plants (duckweeds, azolla, water hyacinth etc.), aquatic animals (snails, clams etc.) and terrestrial-based live feeds (silkworm larvae, maggots etc.), plant processing products (de-oiled cakes and meals, beans, grains and brans) and animal-processing by-products (blood and feather meal, bone meal etc.). Formulated commercial feeds are composed of several ingredients, mixed in various proportions to complement each other and form a nutritionally complete diet. Farm-made and commercial aquafeeds can be fairly easily split depending on whether they are primarily intended for farm use or for commercial sale. Also, while raw materials that are high in moisture and only of local and/or seasonal availability may be used in the preparation of farm-made aquafeeds, commercial fish feed manufacture is predominantly associated with the processing of dry ingredients and the manufacture of a dry product.

The most common aquafeed processing operations can be summarized as raw material size reduction, raw material blending, feed forming, and feed drying. For these processes many and various options of processing equipment are available ranging from simple mortar and pestle to mincers, hammer mills, pelleters, and extruders. Commercial aquafeed manufacture presents special challenges to the traditional feed milling concepts due to the size and variety of animals being cultivated. Moreover, feed for aquatic species requires a higher degree of precision be it the generally finer particle size of ingredients, or the precise mixing of as many as four dozen ingredients into a feed pellet of minute size in comparison to its terrestrial counterpart. These are the compelling reasons why many new aquaculture feed mills are dedicated to produce only aquatic feeds and often employ human food standards in production. Along with the higher standards of production come more expensive and higher quality standards for the ingredients used for what are often very sensitive production animals.

Farm-made Aquafeeds

There is a lack of an exact definition for farm-made aquafeeds. FAO have suggested that farm-made feeds be defined as feeds in pellet or other forms, consisting of one or more artificial and/or natural feedstuffs, produced for the exclusive use of a particular farming activity and not for commercial sale or profit.

PRECAUTIONARY APPROACH TO FISHERY TECHNOLOGY

OBJECTIVE

Recognising that many aquatic resources are overfished and that the fishing capacity presently available jeopardise their conservation and rational use, technological changes aimed solely at further increasing fishing capacity would not generally be seen as desirable.

Instead a precautionary approach to technological changes would aim at:

- Improving the conservation and long-term sustainability of living aquatic resources;
- Preventing irreversible or unacceptable damage to the environment;
- Improving the social and economic benefits derived from fishing, and
- Improving the safety and working conditions of fishery workers.

INTRODUCTION

Fishery technology consists of the equipment and practices used for finding, harvesting, handling, processing and distributing of aquatic resources and their products.

Different fishery technologies will have different effects on the ecosystem, the social structure of fishing communities, the safety of fishery workers and the ease, effectiveness and efficiency of management of the fishery. It is the amount and context in which fishery technology is used (*e. g.* , when, where and by whom) that influence whether the objectives of fisheries management are reached, and not the technology. For instance, the current overfishing of many aquatic resources is the product of both the efficiency of the finding and catching technologies and of the amount used. Similarly, building a fishmeal plant might involuntarily result in severe changes in the way the fishery is conducted, and in the community's social structure.

Fishery technology is constantly evolving and its efficiency in catching fish will increase over time. For example, a 4 per cent increase in efficiency per year would cause a doubling of the fishing mortality rate in 18 years if the fishing effort remained constant. A precautionary approach to management should take such increases into account.

A precautionary approach should be adopted for the development of new technologies or the transfer of existing technologies to other fisheries to avoid unplanned abrupt changes in fishing pressure or social structures. Certain technologies will be considered undesirable, if they create unacceptable effects (*e. g.* , poison or explosives) or if their adoption leads to wasteful use (*e. g.* , at sea, sorting machines have been banned where they might increase discarding).

Fishery technologies produce side effects on the environment and on non-target species. These effects have often been ignored but, in the context of a precautionary approach, some technologies may warrant a review. Similarly, a precautionary approach would encourage careful consideration of the side effects of new fishery technologies before they are introduced.

Each fishery technology has advantages and disadvantages that should be balanced in a precautionary approach, and it may be better to have a mixture of technologies. When new fishery technology is introduced, it should be carefully evaluated to assess its potential direct and indirect effects. If a mix of fishery technology representing "best current practice" in an area can be identified, precautionary management would encourage its adoption while it would discourage damaging ones. Responsible fishery technology achieves the specific fishery management objectives with the minimumdamaging side effects. These concepts (of responsible fishing and best current practices) were addressed by the UN General Assembly and in the Cancun Declaration.

A precautionary approach would provide for a process of initial and on-going review of the effects of fishery technology as it is introduced or evolves in local practice. However, the extent to which a precautionary approach can be applied to the management of technological changes depends on the existing level of management. In some cases, education of fishermen and consumers towards responsible practices may be the only possible approach. Where elaborate research, management and enforcement systems are in place, a wider variety of options are available for application of the precautionary approach. However, although some gears and practices are prohibited they may continue to be used. The adoption of a precautionary approach to the management of new fishery technology depends on the ability to achieve compliance through education and/or enforcement. The following sections assume that institutional arrangements exist to achieve compliance.

EVALUATING THE IMPACTS OF TECHNOLOGIES

A precautionary approach to developing and selecting responsible technologies for fishing requires an appropriate understanding of the consequences of their adoption and use. These consequences, particularly the impacts on non-target species and ecosystems, may be highly uncertain. Nevertheless, some information exists and more can be obtained. The problem of evaluating impacts is relevant both to the use of existing technologies and to the development of new ones, as well as to the introduction of existing technologies to new areas. The description of a given technology would state its relative impacts and advantages for a given species in a specific environment. Target fishery, environmental and ecosystem, socio-economic and legal factors should be considered when evaluating the impacts of fishery technologies.

The factors to consider when evaluating the impacts of fishery technology include:

- Target-fishery factors such as selectivity by size and species (*e. g.* , target, non-target, and protected species; discards; survival of escapees; "Ghost fishing"; and catching capacity);
- Environmental and ecosystem factors such as bio-diversity; habitat degradation; contamination and pollution; generation of debris and rubbish disposal; direct mortality; predator-prey relationships;
- Socio-economic factors such as safety and occupational hazards; training requirements; user conflicts; economic performance; employment; monitoring and enforcement requirements and costs; and techno-economic factors (*i. e.* , infrastructure and service requirements; cost and technological accessibility; product quality; and energy efficiency), and
- Legal factors such as existing legislation; need for new legislation; international agreements; and civil liberties.

These factors could be used to identify beneficial new technologies or damaging ones, to assess the ability of a fishery to accommodate increased use of an established technology and to help direct monitoring and special reporting procedures towards important questions. Technologies for aids to navigation, fish-locating devices, processing and distribution could also be described and evaluated using the above criteria. This will require a suitable description of technologies, cross-referenced against a range of possible impacts. Other elements relevant to the specific technology/area evaluated would also be included.

The approaches used to evaluate impacts will vary according to the human and financial resources available to collect information. If resources are limited, it may be possible to make decisions based on existing information on the impacts of similar technologies in similar environments. Monitoring of existing fishing practices (for example recording of bycatch) will provide additional information. Where financial and human resources are limited, existing information on impacts could be used to do desk studies following the approach to evaluation suggested above. Although some general guidelines can be given, based on known characteristics of types of resources and technology, the most appropriate mix of technologies to be used in a particular fishery should be established on a case-by-case basis, following evaluations made at appropriate regional and national levels. Such evaluations could be refined with practical experience and weighed in accordance with local social and economic values.

In the case of new technologies, or technologies new to an area, pilot studies may be cost-effective in evaluating the impacts and can be useful in demonstrating the benefits of new technology. For example, the introduction of escape ports in lobster traps for undersised individuals demonstrated to fishermen that catch rates of large lobsters increased. On the other hand, pilot studies cannot demonstrate long-term gains such as increased yield per recruit, but they will show the short-term losses.

Considerable resources are required for major experiments to measure effects of fishery technology on the marine environment, but well-designed experiments of this type (either as research projects or via experimental management) will provide the most useful information on which to judge the impacts of technologies in particular areas or habitats. This information may be relevant in other areas than the study sites or fisheries from which the data were derived.

Procedures developed in other contexts for protecting the environment could also be suitable when evaluating new technologies in fisheries or major alterations to existing ones. This would be particularly necessary when there are vulnerable resources or fragile ecosystems, that must be protected. In a precautionary approach, proponents of new fishery technology would be required by the State to provide for a proper evaluation of the potential impacts of new techniques before authorisation is given.

The maximum cost that could be justified for evaluating new fishery technology or practices should be in proportion to the expected benefits and impacts.

IMPLEMENTATION

In a precautionary approach to managing fishery technology, a designated lead authority should have the mandate to evaluate and decide on the acceptability of a proposed new technology, or changes to existing technology, and oversee the impact evaluation procedure. Proponents and other stakeholders should be able to appeal if the proper procedure has not been followed or if the decision by authorities does not appear to agree with the conclusions of the review. As authorisation procedures in the majority of cases would be for minor technical improvements, the procedures could be kept simple and administration costs held at a relatively low level. However, minimal progressive improvements will accumulate over time and periodic reviews of the impacts of existing technology will be necessary. Increases in catching efficiency result from the rapid growth in the use of modern information technologies in most fisheries around the world (acoustic fish detection and identification, gear and vessel monitoring, satellite-based environmental sensing and navigation, and easy inter-vessel communication). However, information, formally treated as "a measure of the reduction of uncertainty", can also potentially improve selectivity, safety and profitability of fishing operations and thus create beneficial effects.

Restricting the use of improved information technologies will rarely be justified or successful and there should be a positive attitude towards technical progress in fisheries in general especially with regards to safety at sea and fishermen's health.

The benefits of technological improvements need adequate extension work and education to encourage their adoption. The promotion of the best technology would benefit from improvement in international cooperation regarding technology transfer, as underscored in UNCED's Agenda 21. The successful international efforts in the Eastern Central Pacific in training crews in effectively avoiding bycatches of dolphins through the use of specifically designed technology is a good example of what can be achieved in this respect.

TECHNOLOGY RESEARCH AND DEVELOPMENT

Fishery technology research in support of a precautionary approach would encourage the improvement of existing technologies and promote the development of appropriate new technologies. Such research would not just concentrate on gears used for capture; for example, research into the cost-effective purification of water supplies to ice plants might considerably reduce post-harvest losses and improve product quality and safety.

Technological developments such as satellite tracking may also help precautionary management by improving monitoring of commercial operations and by enabling research to reduce uncertainty about relevant aspects of fisheries science.

IMPLEMENTATION GUIDELINES

The following measures could be applied in order to implement a precautionary approach to fishery technology development and transfer.

Authority

- Effective mechanisms to ensure that the introduction of technology is subject to review and regulation should be established.

Evaluation Procedures

- A first step in the evaluation procedure is the documentation of the characteristics and amount of the fishery technology currently used.
- Procedures for the evaluation of new technologies with a view to identify their characteristics in order to promote the use of beneficial technologies and prevent usage of those leading to difficult-to-reverse changes should be established.
- These procedures should evaluate with appropriate accuracy the possible impacts of the proposed technology in order to avoid wasteful capital and social investments.
- Authorities should ensure that proponents and other stakeholders understand their obligations and their rights regarding such procedures.
- The extent of the evaluation procedures should match the potential effects of the proposed technology, *e. g.* , from desk study through full scale impact studies, possibly including or leading to pilot projects.

Implementation

- Authorities should implement technology gradually to minimize the risk of irreversible damage or overinvestment.
- Existing technologies and their effect on the environment should be reviewed periodically.
- Technological developments may modify the practices of fishery workers. To achieve the full benefits of the technology and to ensure the safety of fishery workers, training in the proper use of the new technology should be provided.
- In fisheries that are being rehabilitated, the opportunity should be taken to review the mix of technologies used.
- Research into responsible fishery technology should be encouraged.
- Technology research for the reduction of uncertainty in stock assessment and monitoring should be encouraged.

7

Fishing through Net Technology

TECHNIQUES OF FISHING NETS

BOTTOM TRAWLING

Bottom trawling is trawling (towing a trawl, which is a fishing net) along the sea floor. It is also referred to as "dragging". The scientific community divides bottom trawling into benthic trawling and demersal trawling. Benthic trawling is towing a net at the very bottom of the ocean and demersal trawling is towing a net just above the benthic zone.

Bottom trawling can be contrasted with midwater trawling (also known as pelagic trawling), where a net is towed higher in the water column. Midwater trawling catches pelagic fish such as anchovies, tuna, and mackerel, whereas bottom trawling targets both bottom living fish (groundfish) and semi-pelagic species such as cod, squid, shrimp, and rockfish.

Trawling is done by a trawler, which can be a small open boat with only 30 hp (22 kW) or a large factory trawler with 10,000 hp (7,500 kW). Bottom trawling can be carried out by one trawler or by two trawlers fishing cooperatively (pair trawling).

HISTORY

An early reference to fishery conservation measures comes from a complaint about a form of trawling dating from the 14th century, during the reign of Edward III. A petition was presented to Parliament in 1376 calling for the prohibition of a "subtlety contrived instrument called the *wondyrchoum*". This was an early beam trawl with a wooden beam, and consisted of a net 6 m (18 ft) long and 3 m (10 ft) wide,

"of so small a mesh, no manner of fish, however small, entering within it can pass out and is compelled to remain therein and be taken. . . by means of which instrument the fishermen aforesaid take so great abundance of small fish aforesaid, that they know not what to do with them, but feed and fatten the pigs with them, to the great damage of the whole commons of the kingdom, and the destruction of the fisheries in like places, for which they pray remedy. " Another source describes the wondyrchoun as, "three fathom long and ten mens' feet wide, and that it had a beam ten feet long, at the end of which were two frames formed like a colerake, that a leaded rope weighted with a great many stones was fixed on the lower part of the net between the two frames, and that another rope was fixed with nails on the upper part of the beam, so that the fish entering the space between the beam and the lower net were caught. The net had maskes of the length and breadth of two men's thumbs"

The response from the Crown was to "let Commission be made by qualified persons to enquire and certify on the truth of this allegation, and thereon let right be done in the Court of Chancery". Thus, already back in the Middle Ages, basic arguments about three of the most sensitive current issues surrounding trawling - the effect of trawling on the wider environment, the use of small mesh size, and of industrial fishing for animal feed - were already being raised.

Until the late 18th century sailing vessels were only capable of towing small trawls. However, in the closing years of that century a type of vessel emerged that was capable of towing a large trawl, in deeper waters. The development of this type of craft, the sailing trawler, is credited to the fishermen of Brixham in Devon. The new method proved to be far more efficient than traditional long-lining. At first its use was confined to the western half of the English Channel, but as the Brixham men extended their range to the North Sea and Irish Sea it became the norm there too.

By the end of the 19th century there were more than 3,000 sailing trawlers in commission in UK waters and the practice had spread to neighbouring European countries. Despite the availability of steam, trawling under sail continued to be economically efficient, and sailing trawlers continued to be built until the middle of the 1920s. Some were still operating in UK waters until the outbreak of World War II, and in Scandinavia and the Faroe Islands until the 1950s.

English commissions in the 19 century determined that there should be no limitation on trawling. They believed that bottom trawling, like tilling of land, actually increased production. As evidence, they noted that a second trawler would often follow a first trawler, and that the second trawler would often harvest even more fish than the first. The reason for this peculiarity is that the destruction caused by the first trawl resulted in many dead and dying organisms, which temporarily attracted a large number of additional species to feed on this moribund mass.

Bottom trawling has been widely implicated in the population collapse of a variety of fish species, locally and worldwide, including orange roughy, barndoor skate, shark, and many others.

FISHING GEAR

The design requirements of a bottom trawl are relatively simple, a mechanism for keeping the mouth of the net open in horizontal and vertical dimensions, a "body" of net which guides fish inwards, and a "cod-end" of a suitable mesh size, where the fish are collected. The size and design of net used is determined by the species being targeted, the engine power and design of the fishing vessel and locally enforced regulations.

BEAM TRAWLING

The simplest method of bottom trawling, the mouth of the net is held open by a solid metal beam, attached to two "shoes", which are solid metal plates, welded to the ends of the beam, which slide over and disturb the seabed. This method is mainly used on smaller vessels, fishing for flatfish or prawns, relatively close inshore.

OTTER TRAWLING

Otter trawling derives its name from the large rectangular otter boards which are used to keep the mouth of the trawl net open. Otter boards are made of timber or steel and are positioned in such a way that the hydrodynamic forces, acting on them when the net is towed along the seabed, pushes them outwards and prevents the mouth of the net from closing. They also act like a plough, digging up to 15 cm into the seabed, creating a turbid cloud, and scaring fish towards the net mouth. The net is held open vertically on an otter trawl by floats and/or kites attached to the "headline" (the rope which runs along the upper mouth of the net), and weighted "bobbins" attached to the "foot rope" (the rope which runs along the lower mouth of the net). These bobbins vary in their design depending on the roughness of the sea bed which is being fished, varying from small rubber discs for very smooth, sandy ground, to large metal balls, up to 0. 5 m in diameter for very rough ground. These bobbins can also be designed to lift the net off the seabed when they hit an obstacle. These are known as "rock-hopper" gears.

BODY OF THE TRAWL

The body of the trawl is funnel-like, wide at its "mouth" and narrowing towards the cod end, and usually is fitted with wings of netting at the both sides of the mouth. It is long enough to assure adequate flow of water and prevent fish from escaping the net, after having been caught. It is made of diamond-meshed netting, the size of the meshes decreasing from the front of the net towards the codend. Into the body, fish and turtle escape devices can be fitted. These can be simple structures like "square mesh panels", which are easier for smaller fish to pass through, or more complicated devices, such as bycatch grills.

Cod End

The cod end is the trailing end of the net where fish are finally "caught". The size of mesh in the cod end is a determinant of the size of fish which the net catches. Consequently, regulation of mesh size is a common way of managing mortality of juvenile fishes in trawl nets.

HOW TRAWLS WORK

The idea that fish are passively "scooped up" is commonly held, and has been since trawling was first developed, but has been revealed to be erroneous. Since the development of scuba diving equipment and cheap video cameras it has been possible to directly observe the processes that occur when a trawl is towed along the seabed. The trawl doors disturb the sea bed, create a cloud of muddy water which hides the oncoming trawl net and generates a noise which attracts fish.

The fish begin to swim in front of the net mouth. As the trawl continues along the seabed, fish begin to tire and slip backwards into the net. Finally, the fish become exhausted and drop back, into the "cod end" and are caught. The speed that the trawl is towed at depends on the swimming speed of the species which is being targeted and the exact gear that is being used, but for most demersal species, a speed of around 4 knots (7 km/h) is appropriate.

ENVIRONMENTAL IMPACTS

Bottom fishing has operated for over a century on heavily fished grounds such as the North Sea and Grand Banks. While overfishing has long been recognised as causing major ecological changes to the fish community on the Grand Banks, concern has been raised more recently about the damage which benthic trawling inflicts upon seabed communities. A species of particular concern is the slow growing, deep water coral *Lophelia pertusa*. This species is home to a diverse community of deep sea organisms, but is easily damaged by fishing gear.

On November 17, 2004 the United Nations General Assembly urged nations to consider temporary bans on high seas bottom trawling.

Resuspension

Bottom trawling stirs up the sediment at the bottom of the sea. The suspended solid plumes can drift with the current for tens of kilometres from the source of the trawling. These plumes introduce a turbidity which decreases light levels at the bottom and can affect kelp reproduction.

Ocean sediments are the sink for many persistent organic pollutants, usually lipophilic pollutants like DDT, PCB and PAH. Bottom trawling mixes these pollutants into the plankton ecology where they can move back up the food chain and into our food supply.

Phosphorus is often found in high concentration in soft shallow sediments. Resuspending nutrient solids like these can introduce oxygen demand into the water column, and result in oxygen deficient dead zones.

Even in areas where the bottom sediments are ancient, bottom trawling, by reintroducing the sediment into the water column, can create harmful algae blooms. More suspended solids are introduced into the oceans from bottom trawling than any other man-made source.

Deep Sea Impacts

The UN Secretary General reported in 2006 that 95 percent of damage to seamount ecosystems worldwide is caused by deep sea bottom trawling.

CURRENT RESTRICTIONS

Today, some countries regulate bottom trawling within their jurisdictions:

- The United States National Oceanic and Atmospheric Administration banned bottom trawling off most of its Pacific coast in early 2006 and has restricted the practice severely off its other coasts as well. This Federal regulation affects areas between 3–300 miles from the coast (areas within 3 miles (4. 8 km) of the coast are State regulated).
- The Council of the European Union in 2004 applied "a precautionary approach" and closed the sensitive Darwin Mounds off Scotland to bottom trawling.
- In 2005, the FAO's General Fisheries Commission for the Mediterranean (GFCM) banned bottom trawling below 1000 metres and, in January 2006, completely closed ecologically sensitive areas off Italy, Cyprus, and Egypt to all bottom trawling.
- Norway first recognised in 1999 that trawling had caused significant damage to its cold-water lophelia corals. Norway has since established a programme to determine the location of cold-water corals within its EEZ so as to quickly close those areas to bottom trawling.
- Canada has acted to protect vulnerable coral reef ecosystems from bottom trawling off Nova Scotia. The Northeast Channel was protected by a fisheries closure in 2002, and the Gully area was protected by its designation as a Marine Protected Area (MPA) in 2004.
- Australia in 1999 established the Tasmanian Seamounts Marine Reserve to prohibit bottom trawling in the south Tasman Sea. Australia also prohibits bottom trawling in The Great Australian Bight Marine Park near Ceduna off South Australia. In 2004, Australia established the world's largest marine protected area in the Great Barrier Reef Marine Park where fishing and other extractive activities are prohibited.
- New Zealand in 2001 closed 19 seamounts within its EEZ to bottom

trawling, including in the Chatham Rise, sub-Antarctic waters, and off the east and west coasts of the North Island. New Zealand Fisheries Minister Jim Anderton announced on 14 February 2006 that a draft agreement had been reached with fishing companies to ban bottom trawling in 30 percent of New Zealand's exclusive economic zone, an area of about 1. 2 million km^2 reaching from sub-Antarctic waters to sub-tropical ones. But only a small fraction of the area proposed for protection will cover areas actually vulnerable to bottom trawling.

- Palau has banned all bottom trawling within its jurisdiction and by any Palauan or Palauan corporation anywhere in the world.
- The President of Kiribati, Anote Tong, announced in early 2006 the formation of the world's first deep sea marine reserve area. This measure— the Phoenix Islands Protected Area— creates the world's third largest marine protected area and may protect deep sea corals, fish, and seamounts from bottom trawling. However, the actual boundaries of this reserve and what harvest limitations may occur therein have not been detailed. Moreover, Kiribati currently has only 1 patrol boat to monitor this proposed region.

LACK OF REGULATION

Beyond national jurisdictions, most bottom trawling is unregulated either because there is no Regional Fisheries Management Organisation (RFMO) with competence to regulate, or else what RFMOs that do exist have not actually regulated. The major exception to this is in the Antarctic region, where the Convention for the Conservation of Antarctic Marine Living Resources regime has instituted extensive bottom trawling restrictions. The North East Atlantic Fisheries Commission (NEAFC) also recently closed four seamounts and part of the mid-Atlantic Ridge from all fishing, including bottom trawling, for three years. This still leaves most of international waters completely without bottom trawl regulation. As of May 2007 the area managed under the South Pacific Regional Fisheries Management Organisation (SPRFMO) has gained a new level of protection. All countries fishing in the region (accounting for about 25 percent of the global ocean) agreed to exclude bottom trawling on high seas areas where vulnerable ecosystems are likely or known to occur until a specific impact assessment is undertaken and precautionary measures have been are implemented. Also observers will be required on all high seas bottom trawlers to ensure enforcement of the regulations.

FAILED UNITED NATIONS BAN

Palau President Tommy Remengesau has called for a ban on destructive and unregulated bottom trawling beyond national jurisdictions and Palau has led the effort at the United Nations and in the Pacific to achieve a consensus by countries to take this action at an international level. Palau has been joined by the Federated States of Micronesia, the Republic of the Marshall Islands, and Tuvalu in supporting an interim bottom trawling ban at the United Nations. The proposal for this ban did not result in any actual legislation and was blocked. In 2006, New Zealand Fisheries Minister Jim Anderton promised to support a global ban on bottom trawling if there was sufficient support to make that a practical option. Bottom Trawling has been banned in 1/3 of New Zealand's waters (although a large percentage of these areas were not viable for bottom trawling in the first place)

THROW NET

A cast net, also called a throw net, is a net used for fishing. It is a circular net with small weights distributed around its edge. The net is cast or thrown by hand in such a manner that it spreads out on the water and sinks.

Fig. A Fisherman Casting a Net in Kerala, India

This technique is called net casting or net throwing. Fish are caught as the net is hauled back in. This simple device is particularly effective for catching small bait or forage fish, and has been in use, with various modifications, for thousands of years. On the US Gulf Coast, it is used especially to catch mullet, which will not bite a baited hook.

CONSTRUCTION AND TECHNIQUE

Contemporary cast nets have a radius which ranges from 4 to 12 feet (1. 2 to 3. 6 metres). Only strong people can lift the larger nets once they are filled with fish. Standard nets for recreational fishing have a four foot hoop. Weights are usually distributed around the edge at about one pound per foot (1. 5 kilograms per metre). Attached to the net is a landline, one end of which is held in the hand as the net is thrown. When the net is full, a retrieval clamp, which works like a wringer on a mop, closes the net around the fish. The net is then retrieved by pulling on the landline. The net is lifted into a bucket and the clamp is released, dumping the caught fish into the bucket.

Cast nets work best in water no deeper than their radius. Casting is best done in waters free of obstructions. Reeds cause tangles and branches can rip nets. The net caster stands with one hand holding the landline, and with the net draped over the other arm, so the weights dangle. The line is then thrown out to the water, using both hands, in a circular motion rather as in hammer throwing. The net can be cast from a boat, or from the shore, or by wading. There are also optional net throwers that can make casting easier. These look like a lid from a trash can, including the handle on top. The outside circumference has a deep gutter. The net is loaded along the gutter and the weights are placed inside the gutter. The net is then tossed into the water using the thrower.

HISTORY

In Norse mythology the sea giantess Rán cast a fishing net to trap lost sailors. In Ancient Rome, in a parody of fishing, a type of gladiator called a retiarius or "net fighter" was armed with a trident and a cast net. The retiarius was traditionally pitted against a secutor.

Between 177 and 180 the Greek author Oppian wrote the *Halieutica*, a didactic poem about fishing. He described various means of fishing including the use of nets cast from boats. References to cast nets can also be found in the New Testament.

CHEENA VALA

The Chinese fishing nets (Cheena vala) are fishing nets that are fixed land installations for an unusual form of fishing — shore operated lift nets. They are mostly found in the Indian state of Kerala. Huge mechanical contrivances hold out horizontal nets of 20 m or more across. Each structure is at least 10 m high and comprises a cantilever with an outstretched net suspended over the sea and large stones suspended from ropes as counterweights at the other end. Each installation is operated by a team of up to six fishermen.

The system is sufficiently balanced that the weight of a man walking along the main beam is sufficient to cause the net to descend into the sea. The net is left for a short time, possibly just a few minutes, before it is raised by pulling on ropes. The catch is usually modest: a few fish and crustaceans — these may be sold to passers by within minutes.

Rocks, each 30 cm or so in diameter are suspended from ropes of different lengths. As the net is raised, some of the rocks one-by-one come to rest on a platform thereby keeping everything in balance.

Fig. The Chinese Fishing Nets of Fort Cochin.

Each installation has a limited operating depth. Consequently, an individual net cannot be continually operated in tidal waters. Different installations will be operated depending on the state of the tide.

It was earlier thought that the nets might have been introduced by the Chinese explorer Zheng He. Recent research shows that these were introduced by Portuguese Casado settlers from Macau

The Chinese fishing nets have become a very popular tourist attraction, their size and elegant construction is photogenic and the slow rhythm of their operation is quite hypnotic. In addition, catches can be purchased individually and need be taken only a short distance to a street entrepreneur who will cook it. Similar style fishing nets are also seen in Vietnam.

LAMPUKI NETTING

Lampuki is the Maltese name for the dorado or mahi-mahi, a kind of fish that migrates past the Maltese islands during the autumn. The fishing season for lampuki is from the end of August through to November.

Fishermen cut and gather the larger, lower fronds from palm trees which they then weave into large flat rafts. The rafts are pulled out to sea, usually with the small traditional fishing boats known as Luzzu,but can also be pulled out to sea by using larger modern fishing boats. During midday lampuki school underneath the rafts, seeking the shade. The fishermen will first stay 5-10 metres away from the raft and will repeatedly go round the raft pulling a silicone squid jig behind them till they catch something, after they catch something they will leave the lampuka (dorado,mahi-mahi) hanging to the side of the boat till the other fish will come and school next to it then a mesh net is thrown over the schooling fish. This method is known as kannizzati and has not changed significantly since Roman times. The lampuki are used both for local consumption as well as export.

CORACLE BOTING TECHNIQUES

The coracle is a small, lightweight boat of the sort traditionally used in Wales but also in parts of Western and South Western England, Ireland (particularly the River Boyne), and Scotland (particularly the River Spey); the word is also used of similar boats found in India, Vietnam, Iraq and Tibet. The word "coracle" comes from the Welsh *cwrwgl*, cognate with Irish and Scottish Gaelic *currach*, and is recorded in English as early as the sixteenth century. Other historical English spellings include *corougle*, *corracle*, *curricle* and *coricle*.

STRUCTURE

Oval in shape and very similar to half a walnut shell, the structure is made of a framework of split and interwoven willow rods, tied with willow bark. The outer layer was originally an animal skin such as horse or bullock hide (corium), with a thin layer of tar to make it fully water proof – today replaced by tarred calico or canvas, or simply fibreglass. The Vietnamese/Asian version of the coracle is made somewhat differently: using interwoven bamboo and waterproofed by using resin and coconut oil. The structure has a keel-less, flat bottom to evenly spread the weight of the boat and its load across the structure and to reduce the required depth of water — often to only a few inches, making it ideal for use on rivers.

Each coracle is unique in design, as it is tailored to the river conditions where it was built and intended to be used. In general there is one design per river, but this is not always the case.

The Teifi coracle, for instance, is flat bottomed, as it is designed to negotiate shallow rapids, common on the river in the summer, while the Carmarthen coracle is rounder and deeper, because it is used in tidal waters on the Tywi, where there are no rapids. Teifi coracles are made from locally harvested wood — willow for the lats (body of the boat), hazel for the weave (Y bleth in Welsh — the bit round the top) — while Tywi coracles have been made from sawn ash for a long time. The working boats tend to be made from fibreglass these days. Teifi coracles use no nails, relying on the interweaving of the lats for structural coherence, whilst the Carmarthen ones use copper nails and no interweaving.

They are an effective fishing vessel because, when powered by a skilled man, they hardly disturb the water or the fish, and they can be easily manoeuvred with one arm, while the other arm tends to the net; two coracles to a net. The coracle is propelled by means of a broad-bladed paddle, which traditionally varies in design between different rivers. It is used in a sculling action, the blade describing a figure-of-eight pattern in the water. The paddle is used towards the front of the coracle, pulling the boat forward, with the paddler facing in the direction of travel. Another important aspect to the Welsh Coracle is that it can be carried on his back by one man. 'Llwyth dyn ei gorwgl' — the load of a man is his coracle.

HISTORY

Designed for use in the swiftly flowing streams of Wales and parts of the rest of Britain and Ireland, the coracle has been in use for centuries, having been noted by Julius Caesar in his invasion of Britain in the mid first century BC, and used in his campaigns in Spain. Remains interpreted as a possible coracle were found in a Bronze Age grave from near Dalgety Bay, and two others have been described, from Corbridge and from near Ferriby.

According to Ian Harries, coracle fisherman, coracles are so light and portable that they can easily be carried on the fisherman's shoulders when proceeding to and from his work. Coracle fishing is performed by two coraclers. Where fishing is performed by two people, there is one fisherman per coracle. The net is stretched across the river between the two coracles (the coracler will paddle one handed, dragging the net in the other) and drawn downstream. When a fish is caught, each hauls up an end of the net until the two coracles are brought to touch, and the fish is then secured, using a priest (or knocker – a small block of wood) to stun the fish.

TODAY

Coracles are now seen regularly only in tourist areas of West Wales, and irregularly in Shropshire on the River Severn – a public house in Sundorne, Shrewsbury called "The Coracle" has a pub sign featuring a man using a coracle on a river. The Welsh Rivers Teifi and Tywi are the best places to find coracles in Wales, although the type of coracle differs depending on the river. On the Teifi they are most frequently seen between Cenarth, and Cilgerran and the village of Llechryd.

In 1974 a Welsh coracle piloted by Bernard Thomas of Llechryd crossed the English Channel to France in 13½ hours. The journey was undertaken to demonstrate how the Bull Boats of the Mandan Indians of North Dakota could have been copied from coracles introduced by Prince Madog in the 12th century.

For many years until 1979, Shrewsbury coracle maker Fred Davies achieved some notability amongst football fans; he would sit in his coracle during Shrewsbury Town FC home matches at Gay Meadow, and retrieve stray balls from the River Severn. Although Davies died in 1994, his legend is still associated with the club.

SAFETY

The design of the craft, as explained above, makes the coracle an unstable craft. Because it sits "on" the water, rather than "in" it, they can easily be carried by currents and the wind. The Coracle Society has published guidelines for safely using coracles.

SIMILAR CRAFT

Fig. The Ku-Dru or Kowa of Tibet is Very Similar to a Coracle

The earliest known written evidence of a coracle-type boat (quffa), still in use today, is in the Bible, Exodus 2:3.

The Irish curach (also *currach* or *curragh*) is a similar, but larger, vessel still in use today. Curachs were also used in the west of Scotland:

"The *curach* or boat of leather and wicker may seem to moderns a very unsafe vehicle, to trust to tempestuous seas, yet our forefathers fearlessly committed themselves in these slight vehicles to the mercy of the most violent weather. They were once much in use in the Western Isles of Scotland, and are still found in Wales. The framework [in Gaelic] is called *crannghail*, a word now used in Uist to signify a frail boat. " *Dwelly's [Scottish] Gaelic Dictionary*[13]

The currachs in the River Spey were particularly similar to Welsh coracles. Other related craft include:

- India – *parisal*
- Iraq – *gufa* or *quffa*
- Native American societies – *bull boat*
- Tibet – *ku-dru* and *kowas*
- Vietnam – *thung-chai*

FISHING TECHNIQUE THROUGH DRIFT NETTING

Drift netting is a fishing technique where nets, called drift nets, are allowed to float freely at the surface of a sea or lake. Usually a drift net is a gill net with floats attached to a rope along the top of the net, and weights attached to another rope along the foot of the net to keep it vertical in the water.

Drift nets are placed by ships and are left free-floating until retrieved. These nets usually target schools of pelagic fish. Drift nets are a type of gill net because of the tendency for the fishes' gills to get caught in the net.

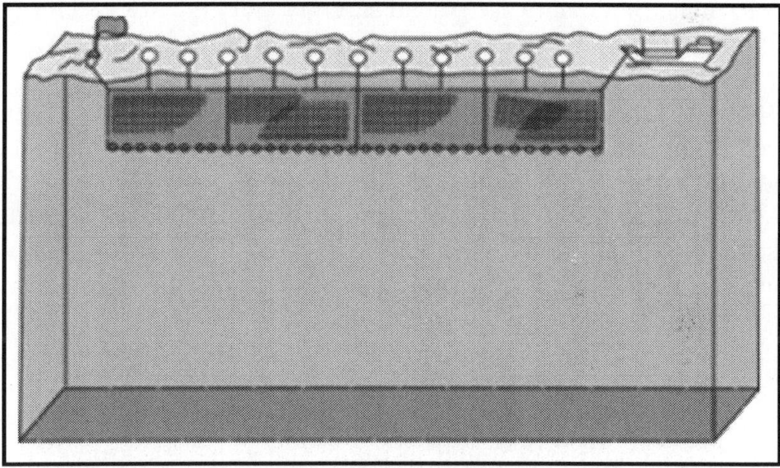

Fig. Drift Netting.

Traditionally drift nets were made of organic materials, such as hemp, which were biodegradable. Prior to 1950, nets tended to have a larger mesh size. The larger mesh only caught the larger fish, allowing the smaller, younger ones to slip through. When drift net fishing grew in scale during the 1950s, the industry changed to synthetic materials with smaller mesh size. Synthetic nets last longer, are odourless and nearly invisible in the water, and do not biodegrade.

Drift net fishing became a commercial fishing practice because it is cost effective. Nets can be placed by low-powered vessels making it fuel efficient. Drift nets are also effective at bringing in large amounts of fish in one catch.

Prior to the 1960s net size was not limited, and commercially produced nets were commonly as long as 50 kilometres (31 mi). In 1987 the U. S. , enacted the Driftnet Impact, Monitoring, Assessment and Control Act limiting the length of nets used in American waters to 1. 5 nautical miles (~1. 7 miles). In 1989 the United Nations General Assembly (UNGA) placed a moratorium on the practice of drift net fishing. In 1992 the UN banned the use of drift nets longer than 2. 5 km long in international waters.

CONTROVERSY

By-catch

Any fish that crosses the path of a drift net in the ocean may be tangled or caught in the net. Non-target individuals caught in the net are called by-catch. In 1994 United Nations Food and Agriculture Organisation (UNFAO) estimated global by-catch rates to be as high as 27 million tons of fish discarded by fisheries each year. Many individuals of non-target species perish as by-catch in the cast of each drift net. As a result, many such species are now endangered. Species caught as by-catch include sharks, dolphins, whales, turtles, sea birds, and other marine mammals. Since nets are placed and may not be retrieved for days, air-breathing mammals that become tangled in the nets drown if they are unable to free themselves.

In 1990s drift net fisheries were responsible for 30,000 tons of sharks and skates in global by-catch annually. While filming National Geographic's *Incidental Kill* in the California Channel Islands where swordfish and various sharks swim north, the divers discovered that many drift net boats had placed nets that night. The nets were one mile long each and nearly 100 feet (30 m) high placed to target swordfish and thresher sharks. They swam half the length of one net and in that length discovered 32 dead blue sharks in the net as well as 2 hammer head sharks, a sea lion, and a manta ray all of which were thrown back into the ocean when the net was hauled in.

Although long line fisheries are the main contributor to sea bird by-catch, sea birds are also caught in drift nets in significant numbers. Studies conducted on 30 small-scale drift net fisheries in the Baltic Sea estimate that 90,000 sea birds die annually in drift nets.

Bycatch is thrown back to the ocean either dead or with injuries that may result in death. If not eaten, dead animals decompose, as bacteria use oxygen to break down the organic matter. Large amounts of dead matter decomposing in the ocean causes the surrounding levels of dissolved oxygen to decrease.

Environmental Damage

Drift nets lost or abandoned at sea due to storms causing strong currents, accidental loss, or purposeful discard become ghost nets. Synthetic nets are resistant to rot or breakdown, therefore ghost nets fish indefinitely in the oceans. Marine animals are easily tangled in ghost nets. The float line on the net allows it to be pushed in the current which causes ecological damage to plant life and substrate habitats as the nets drag the sea floor.

Illegal Fishing

Each country has jurisdiction over the waters within 200 nautical miles of their shores, called the exclusive economic zone, set by the Law of the Sea. Outside these boundaries lie international waters, or the high seas. While fishing in international waters, vessels must comply with regulations of the country in whose flag they fly, but there are no enforcers on the high seas. International waters make up 50 per cent of the world's surface, yet are its least protected habitat.

Declining fish stocks have caused illegal fishing practices to increase. Illegal, unregulated, or unreported fishing catch between 11 and 26 million tons a year which accounts for one quarter of global catch.

Illegal fishing includes taking undersized fish, fishing in closed waters, taking more fish than permitted, or fishing during seasonal closures. Illegal fishing is prominent due to lack of enforcement or punishments.

Despite controls, violations of drift net fishing laws are commonplace. The Mediterranean Sea is the most overexploited. With 21 modern states with coastline on the sea, there are many fisheries harvesting one small area. When drift net gear was banned, manufacturers modified the design of the nets so they no longer fell under the that definition.

A new definition was established in 2007 as "any gillnet held on the sea surface or at a certain distance below it by floating devices, drifting with the current, either independently or with the boat to which it may be attached. It may be equipped with devices aiming to stabilize the net or to limit drift".

Japanese drift net fishing began to draw public attention in the mid-1980s when Japan and other Asian countries began to send large fleets to the North Pacific Ocean to catch tuna and squid. Japan operated about 900 drift net vessels, earning around $300 million a year.

Those fishing boats were blamed not only for indiscriminate destruction of marine life, but also for poaching North Pacific salmon, harming the U. S. , and Canadian fishing industries, and threatening the jobs of fishermen who did not use such methods. The first Bush administration opposed a U. S. , driftnet ban because it would allegedly conflict with a treaty with Japan and Canada regarding salmon fishing in the North Pacific.

OTHER USES

Drift nets also are used in ecological studies in regard to downstream drift of invertebrates. The nets are strung across a stream and allowed to sit over night, collecting samples. These nets are crucial in the understanding of how watersheds function.

FISHING METHOD USED BY GILLNETTING TECHNIQUES

Gillnetting is a common fishing method used by commercial and artisanal fishermen of all the oceans and in some freshwater and estuary areas. "Gill nets are vertical panels of netting normally set in a straight line.

Fish may be caught by gill nets in 3 ways:

1. *Wedged:* Held by the mesh around the body
2. *Gilled:* Held by mesh slipping behind the opercula, or
3. *Tangled:* Held by teeth, spines, maxillaries, or other protrusions without the body penetrating the mesh. Most often fish are gilled. A fish swims into a net and passes only part way through the mesh. When it struggles to free itself, the twine slips behind the gill cover and prevents escape. "

Fig. Oil Painting of Gillnetting.

Gillnets are so effective that their use is closely monitored and regulated by fisheries management and enforcement agencies. Mesh size, twine strength, as well as net length and depth are all closely regulated to reduce bycatch of non-target species. Gillnets have a high degree of size selectivity Most salmon fisheries in particular have an extremely low incidence of catching non-target species.

LEGAL STATUS

United Nations General Assembly Resolution 46/215 called for the cessation of all "large-scale pelagic drift-net fishing" in international waters by the end of 1992. The laws of individual countries vary with regard to fishing in waters under their jurisdiction.

Possession of gillnets is illegal in some U. S. , states. Oregon voters had the chance to decide on whether gillnetting will continue in the Columbia river in November, 2012 by voting on Measure 81. [16] The measure was defeated with 65 per cent of Oregon voters voting against the measure and allowing commercial gillnet fishing to continue on the Columbia River.

SELECTIVITY

Gillnets are a series of panels of meshes with a weighted "foot rope" along the bottom, and a *headline*, to which floats are attached. By altering the ratio of floats to weights, buoyancy changes, and the net can therefore be set to fish at any depth in the water column. In commercial fisheries, the meshes of a gillnet are uniform in size and shape. Fish smaller than the mesh of the net pass through unhindered, while those too large to push their heads through the meshes as far as their gills are not retained. This gives gillnets the ability to target a specific size of fish, unlike other net gears such as trawls, in which smaller fish pass through the meshes and all larger fish are captured in the net.

Commercial gillnet fisheries are still an important method of harvesting salmon in Alaska, British Columbia, Washington, and Oregon. In the lower Columbia River, non-Indian commercial salmon fisheries for spring Chinook have developed methods of selectively harvesting adipose fin clipped hatchery salmon using small mesh gillnets known as tangle nets or tooth nets. Non-adipose fin clipped fish (primarily natural origin salmon) must be released. Fishery management agencies estimate a relatively low release mortality rate on salmon and steelhead released from these small mesh gillnets. Gillnets are sometimes a controversial gear type especially among sport fishers who argue they are inappropriate especially for salmon fisheries. These arguments are often related to allocation issues between commercial and recreational (sport) fisheries and not conservation issues. Most salmon fisheries, especially those targeting Pacific salmon in North America, are strictly managed to minimize total impacts to specific populations and salmon fishery managers continue to allow the use of gillnets in these fisheries.

In 2012, Stephen Mathews from the University of Washington compared Puget Sound bycatch data for the non-treaty gillnet and purse seine keta salmon fisheries. He found that although neither fishery had major bycatch problems with non-target salmonids, the gillnet fishery has substantially less impact on non-target Chinook salmon.

ALTERNATIVES

Given the selective properties of gillnet fishing, alternative methods of harvest are currently being studied. Recent WDF&W reports suggest that purse seine is the most productive method with having highest catch per unit effort (CPUE), but has little information on the effectiveness of selectively harvesting hatchery-reared salmon. More conclusive research has been conducted jointly between the Confederated Tribes of the Colville Reservation and Bonneville Power Administration on a 10-year study on selective harvest methods of hatchery origin salmon in the Upper Columbia River by purse seine and tangle net. Their 2009 and 2010 findings show that purse seines have a higher percentage of survivability and higher CPUE than does tangle nets. A Colville Tribe biologist reports that during these two years the tribe harvested 3,163 hatchery Chinook while releasing 2,346 wild Chinook with only 1. 4 per cent direct or immediate mortality using purse seines, whereas the tangle net was far less productive but had an approximate 12. 5 per cent mortality. Researchers commented that the use of recovery boxes and shortened periods between checking the nets would have likely decreased mortality rates. While there is data that shows success of selective methods of harvest at protecting wild and ESA listed salmon, there still must be social acceptance of new methods of fishing.

TYPES OF GILLNETS

The FAO classifies gillnet gear types as follows:

Set Gillnets

Set gillnets consist of a single netting wall kept vertical by a floatline (upper line/headrope) and a weighted groundline (lower line/footrope). Small floats, usually shaped like eggs or cylinders and made of solid plastic, are evenly distributed along the floatline, while lead weights are evenly distributed along groundline.

The lower line can also be made of lead cored rope, which does not need additional weight. The net is set on the bottom, or at a distance above it and held in place with anchors or weights on both ends. By adjusting the design these nets can fish in surface layers, in mid water or at the bottom, targeting pelagic, demersal or benthic species. On small boats gillnets are handled by hand. Larger boats use hydraulic net haulers or net drums. Set gillnets are widely used all over the world, and are employed both in inland and sea waters. They are popular with artisanal fisheries because no specialised gear is needed, and it is low cost based on the relationship of fuel/fish.

Encircling Gillnets

Encircling gillnets are gillnets set vertically in shallow water, with the floatline remaining at the surface so they encircle fish. Small open boats or canoes can be used to set the net around the fish. Once the fish are encircled, the fishers shout and splash the water to panic the fish so they gill or entangle themselves. There is little negative impact on the environment. As soon as the gear is set the scaring takes place and the net is hauled back in. The fish are alive and discards can be returned to the sea. Encircling gillnets are commonly used by groups of small-scale fishers, and does not require other equipment.

Fig. A Gill Netter

Combined Gillnets-trammel Nets

This bottom-set gear has two parts:

- The upper part is a standard gillnet where semi-demersal or pelagic fish can be gilled
- The lower part is a trammel net where bottom fish can entangle.

The combined nets are maintained more or less vertically in the usual way by floats on the floatline and weights on the groundline. They are set on the bottom. After a time depending on the target species, they are hauled on board. Traditional combined nets were hauled by hand, especially on smaller boats. Recent hydraulic driven net haulers are now common. The gilled, entangled and enmeshed fish are removed from the net by hand. Of some concern with this method is ghost fishing by lost nets and bycatch of diving seabirds. Nets combined in this way were first used in the Mediterranean.

DRIFT NETS

A drift net consists of one or more panels of webbing fastened together. They are left free to drift with the current, usually near the surface or not far below it. Floats on the floatline and weights on the groundline keep them vertical. Drift nets drift with the current while they are connected with the operating vessel, the driftnetter or drifter.

Drift nets are usually used to catch schooling forage fish such as herring and sardines, and also larger pelagic fish such as tuna, salmon and pelagic squid. Net haulers are usually used to set and haul driftnets, with a drifter capstan on the forepart of the vessel. In developing countries most nets are hauled by hand. The mesh size of the gillnets is very effective at selecting or regulating the size of fish caught. The drift net has a low fuel/fish energy consumption compared to other fishing gear. However, the issue of concern with this type of net is the bycatch of species that are not targeted, such as marine mammals, seabirds and to a minor extent turtles. The use of drift nets longer than 2. 5 kilometres on the high seas was banned by the United Nations in 1991. Prior to this ban, drift nets were reaching lengths of 60 kilometres. However, there are still serious concerns with ongoing violations.

GILLNETS AND ENTANGLING NETS

The tangle net, or tooth net, originated in British Columbia, Canada, as a gear specifically developed for selective fisheries. Tangle nets have smaller mesh sizes than standard gillnets. They are designed to catch fish by their nose or jaw, enabling bycatch to be resuscitated and released unharmed. Tangle nets as adapted to the mark-selective fishery for spring Chinook salmon on the lower Columbia River have a standard mesh size of 4-1/4 inches (10. 8 cm.). Short net lengths and soak times are used in an effort to land fish in good condition. Tangle nets are typically used in situations where the release of certain (usually wild) fish unharmed is desirable. In a typical situation calling for the use of a tangle net, for instance, all fish retaining their adipose fins (usually wild) must be returned to the water. Tangle nets are used in conjunction with a live recovery box, which acts as a resuscitation chamber for unmarked fish that appear lethargic or stressed before their release into the water.

FISHING GHOST NET

Fig. Sea Turtle Entangled in a Ghost Net

Ghost nets are fishing nets that have been left or lost in the ocean by fishermen. These nets, often nearly invisible in the dim light, can be left tangled on a rocky reef or drifting in the open sea. They can entangle fish, dolphins, sea turtles, sharks, dugongs, crocodiles, seabirds, crabs, and other creatures, including the occasional human diver. Acting as designed, the nets restrict movement, causing starvation, laceration and infection, and suffocation in those that need to return to the surface to breathe.

DESCRIPTION

Some commercial fisherman use gillnets. These are suspended in the sea by flotation buoys, such as glass floats, along one edge. In this way they can form a vertical wall hundreds of metres long, where any fish within a certain size range can be caught. Normally these nets are collected by fishermen and the catch removed. However if this is not done the net can continue to catch fish until the weight of the catch exceeds the buoyancy of the floats. The net then sinks, and the fish are devoured by bottom-dwelling crustaceans and other fish. Then the floats pull the net up again and the cycle continues. Given the high-quality synthetics that are used today, the destruction can continue for a long time. The problem is not just nets; old-fashioned crab pots, without the required "rot-out panel", also sit on the bottom, where they become self-baiting traps that go on catching crabs year after year. Even balled-up fishing line can be deadly for a variety of creatures, including birds and marine mammals. Over time the nets become more and more tangled. In general, fish are less likely to be trapped in gear that has been down a long time.

The French government offered a reward for ghost nets handed in to local coastguards along sections of the Normandy coast between 1980 and 1981. The project was abandoned when people vandalised nets to claim rewards, without retrieving anything at all from the shoreline or ocean.

SCOOP NET

A hand net, also called a scoop net, is a net or mesh basket held open by a hoop. It may or may not be on the end of a handle. Hand nets have been used since antiquity and can be used for scooping fish near the surface of the water, such as muskellunge or northern pike.

Fig. Fishing for Salmon with a Hand or Dip Net on the Fraser River, Canada

A hand net with a long handle is often called a dip net. There are popular contemporary dip net Sockeye Salmon fisheries in Chitina, Kenai River,and Kasilof River Alaska, typically lasting two to three weeks, and is regarded as a subsistence fishery for Alaska residents only. Dip nets can also be used to scoop crabs in shallow water. The basket is made of wire or nylon mesh, rather than cloth mesh, since crabs fight, bite, twist and turn when they are caught.

When a hand net is used by an angler to help land a fish it is called a landing net. Because hand netting is not destructive to fish, hand nets are often used for tag and release, or to capture aquarium fish.

HISTORY

Hand nets have been widely used by traditional fishermen. Small fish are caught both in the shallow water of lagoons and in the open sea. They are made in different sizes ranging from small nets held in one hand to large scoop nets worked by several men. Historically, the Karuk people of the upper Klamath River harvested fish with dip nets.

In England, hand netting is the only legal way of catching eels and has been practised for thousands of years on the River Parrett and River Severn.

MIDWATER TRAWLINGNET FISHING

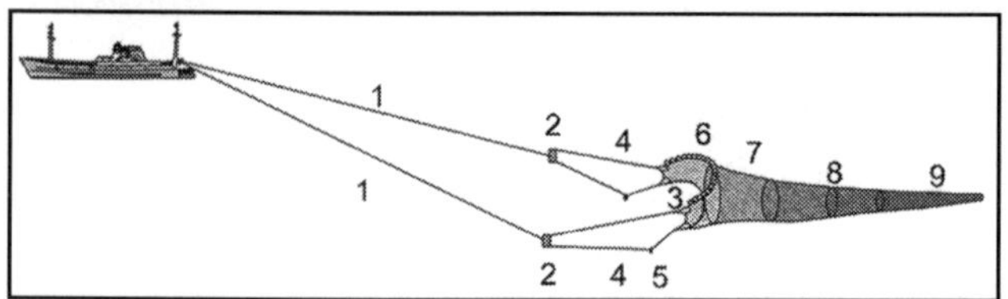

Fig. Midwater (Pelagic) Otter Trawl

1. Trawl warp,
2. Otter boards,
3. Longline chains,
4. Hunter,
5. Weights
6. Headline with floats,
7. Pre-net,
8. Tunnel and belly,
9. Codend

Midwater trawling is trawling, or net fishing, at a depth that is higher in the water column than the bottom of the ocean. It is contrasted with bottom trawling. Midwater trawling is also known as pelagic trawling and bottom trawling as benthic trawling.

In midwater trawling, a cone-shaped net can be towed behind a single boat and spread by trawl doors, or it can be towed behind two boats (pair trawling) which act as the spreading device. Midwater trawling catches pelagic fish such as anchovies, shrimp, tuna and mackerel, whereas bottom trawling targets both bottom living fish (groundfish) and semi-pelagic fish such as: cod, squid, halibut and rockfish.

Whereas bottom trawling can leave serious incidental damage to the sea bottom in its trail, midwater trawling by contrast is relatively benign.

LAVE FISHING NET USED IN RIVER

A lave net is a type of fishing net used in river estuaries, particularly in the Severn Estuary in Wales and England to catch salmon.

Fig. Fisherman Carrying a Lave Net

The lave net is a "Y" shaped structure consisting of two arms called *rimes* made from willow, which act as a frame work to the loosely hung net. The handle is called the *rock staff* and is made of ash or willow. The arms are hinged to the rock staff and are kept in position while fishing with a wooden spreader called the *headboard*. Fishermen wade out at low tide with lave nets on their shoulders to the fishing grounds, with the water up to their waists. The net is then opened and lowered into the outgoing tide which rushes through the net. With his fingers placed at the bottom meshes of the net, the fisherman then waits for the fish to hit the net. The last lave net fishermen in Wales promote the fishery as a tourist attraction at Black Rock, Portskewett, with the aim of maintaining its history and tradition. Demonstrations of lave net fishing can be watched on certain days from the picnic site at Black Rock.

On the English side of the Severn, lave net fishing was practised for centuries at Oldbury on Severn. In the 1990s the fishery declined because the fishing stations silted up, claimed by the fishermen to be a result of slower tides caused by the construction of the Second Severn Crossing. In the past, sturgeon have also been caught in lave nets.

SEINE METHOD OF FISHING

Fig. A basic Seine Net

Seine fishing (or seine-haul fishing) is a method of fishing that employs a seine or dragnet. A seine is a fishing net that hangs vertically in the water with its bottom edge held down by weights and its top edge buoyed by floats. Seine nets can be deployed from the shore as a beach seine, or from a boat.

Seines have been used widely in the past, including by stone age societies. For example, with the help of large canoes, pre-European Maori deployed seine nets which could be over one thousand metres long. The nets were woven from green flax, with stone weights and light wood or gourd floats, and could require hundreds of men to haul. American Native Indians on the Columbia River wove seine nets from spruce root fibres or wild grass, again using stones as weights. For floats they used sticks made of cedar which moved in a way which frightened the fish and helped keep them together. Seine net are also well documented in antiquity. They appear in Egyptian tomb paintings from 3000 BC. In ancient Greek literature, Ovid makes many references to seine nets, including the use of cork floats and lead weights.

Boats deploying seine nets are known as seiners. There are two main types of seine net deployed from seiners: *purse seines* and *Danish seine*

PURSE SEINE

A common type of seine is a purse seine, named such because along the bottom are a number of rings. A line (referred to as a purse-line) passes through all the rings, and when pulled, draws the rings close to one another, preventing the fish from "sounding", or swimming down to escape the net. This operation is similar to a traditional style purse, which has a drawstring. The purse seine is a preferred technique for capturing fish species which school, or aggregate, close to the surface: such as sardines, mackerel, anchovies, herring, certain species of tuna (schooling); and salmon soon before they swim up rivers and streams to spawn (aggregation). Boats equipped with purse seines are called purse seiners.

POWER BLOCK

The power block is a mechanised pulley used on some seiners to haul in the nets. According to the UN Food and Agriculture Organisation, no single invention has contributed more to the success of purse seine net hauling than the power block.

The Puretic power block line was introduced in the 1950s and was the key factor in the mechanisation of purse seining. The combination of these blocks with advances in fluid hydraulics and the new large synthetic nets changed the character of purse seine fishing. The original Puretic power block was driven by an endless rope from the warping head of a winch. Nowadays, power blocks are usually driven by hydraulic pumps powered by the main or auxiliary engine. Their rpm, pull and direction can be controlled remotely.

A minimum of three people are required for power block seining; the skipper, skiff operator, and corkline stacker. In many operations a fourth person stacks the leadline, and oftentimes a fifth person stacks the web.

DRUM

In Canada, specifically on the coast of British Columbia drum seining is a method of seine fishing which was adopted in the late 1950s and is now used exclusively in that region.

The drum seine uses a horizonally mounted drum to haul and store the net instead of a power block. The net is pulled in over a roller, which spans the stern, and then passes through a spooling gear with upright rollers. The spooling gear is moved from side to side across the stern which allows the net to be guided and wound tightly on the drum.

There are several advantages to the drum seine over the power block. The net can be hauled very quickly, more than twice the speed of using a power block, the net does not require overhead handling and the process is therefore safer. The most important advantage is that the drum system can be operated with fewer deckhands.

DANISH SEINE

A Danish seine, also occasionally called an anchor seine, consists of a conical net with two long wings with a bag where the fish collect. Drag lines extend from the wings, and are long so they can surround an area.

A Danish seine is similar to a small trawl net, but the wire warps are much longer and there are no otter boards. The seine boat drags the warps and the net in a circle around the fish. The motion of the warps herds the fish into the central net.

Danish seiner vessels are usually larger than purse seiners, though they are often accompanied by a smaller vessel. The drag lines are often stored on drums or coiled onto the deck by a coiling machine. A brightly coloured buoy, anchored as a "marker", serves as a fixed point when hauling the seine. A power block, usually mounted on a boom or a slewing deck crane, hauls the seine net.

Danish seining works best on demersal fish which are either scattered on or close to the bottom of the sea, or are aggregated (schooling). They are used when there are flat but rough seabeds which are not trawlable. It is especially useful in northern regions, but not much in tropical to sub-tropical areas.

The net is deployed, with one end attached to an anchored dan (marker) buoy, by the main vessel, the seiner, or by a smaller auxiliary boat. A drag line is paid out, followed by a net wing. As the seiner sweeps in a big circle returning to the buoy, the deployment continues with the seine bag and the remaining wing, finishing with the remaining drag line. In this way a large area can be surrounded. Next the drag lines are hauled in using rope-coiling machines until the catch bag can be secured. The seine netting method developed in Denmark. Scottish seining ("fly dragging") was a later modification. The original procedure is much the same as fly dragging except for the use of an anchored marker buoy when hauling, and closing the net and warps and net by winch.

FISHING THROUGH SURROUNDING NET

Fig. The Purse Seine is an example of a Surrounding Net. Here any Salmon Swimming near the Surface are Surrounded with a Wall of Netting, Supported by Floats.

A surrounding net is a fishing net which surrounds fish on the sides and underneath. It is typically used by commercial fishers, and pulled along the surface of the water. There is typically a purse line at the bottom, which is closed when the net is hauled in.

TOOTH NET

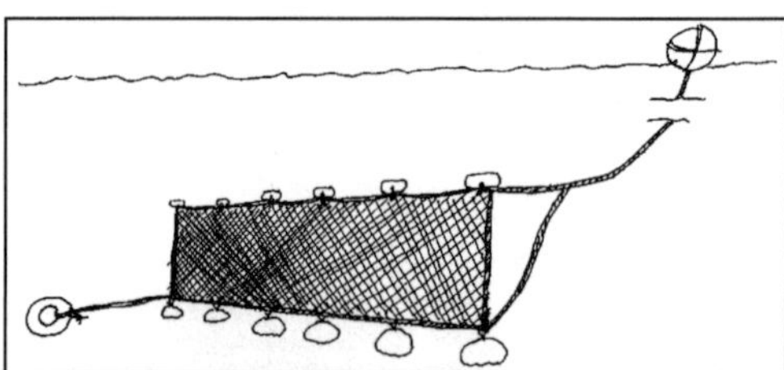

Fig. Diagram of a Tangle Net Shown Upright for Viewing. When used by Fishermen in the Philippines it is Positioned on the Sea Bottom against a Near-vertical Underwater Cliff Wall (drop-off zone).

Similar to a gillnet, the tangle net, or tooth net, is a type of nylon fishing net. Left in the water for no more than two days, and allowing bycatch to be released alive, this net is considered to be less harmful that other nets. The tangle net is used in the Philippines by commercial fishermen, as well as by the scientific community. When spent, these nets can be bundled, and left on the sea floor to collect smaller species. These bundles are known locally as lumen lumen nets.

DESCRIPTION AND TECHNIQUE

The tangle net originated in British Columbia, Canada, as a gear specifically developed for selective fisheries. Tangle nets have smaller mesh sizes than standard gillnets. They are designed to catch fish by their nose or jaw, enabling bycatch to be resuscitated and released unharmed. These nets are made with a very thin light nylon rope, have a small mesh and are strung between two ropes, a top rope with floats, and a bottom rope with weights. Dropped to the bottom of the ocean, located and retrieved through the use of a guide line and buoy, these nets have allowed both fishermen and scientists to reach areas not previously accessible. Tangle nets are generally left on the bottom for no more than a day or two so that the fish and bycatch does not die and spoil.

USE

For the last two decades tangle nets have been set in deep water near steep underwater cliffs off many islands in the Philippines by local fishermen in order to supplement their income through catching commercially valuable mollusks. Scientists have used this technique in recent years to explore the deep water marine habitat. The rich species diversity of the Philippine Islands has been explored through the use of tangle nets which are able to obtain specimens from areas not reachable by traditional methods of using trawls and dredges. Through the placement of tangle nets 50 to 100 meters long, at depths from 100 meters to 400 meters, this cryptic marine habitat has been explored and many new and/or rare species of gastropods and crustaceans have been acquired. The success of deep set tangle nets was further exploited when Philippe Bouchet and Danilo Largo launched an expedition in 2004 to explore the Panglao area, known as the 2004 Panglao Marine Biodiversity Project.

LUMEN LUMEN NETS

When tangle nets are damaged beyond repair they are twisted and wrapped into long bundles, which the Philippine locals call lumen lumen nets. These long, sausage-like bundles are placed on the sea bottom along drop-offs in deep water with strong currents and left for several months at a time, which allows time for veligers to settle and larvae to grow. Lumen lumen nets have yielded many more species of marine animals, including many very small species of micromollusks.

TRAWLING METHOD OF FISHING

Trawling is a method of fishing that involves pulling a fishing net through the water behind one or more boats. The net that is used for trawling is called a trawl.

The boats that are used for trawling are called trawlers or draggers. Trawlers vary in size; from small open boats with only 30 hp engines to large factory trawlers with over 10,000 hp. Trawling can be carried out by one trawler or by two trawlers fishing cooperatively (pair trawling).

Fig. Trawl Net with Fish

Trawling can be contrasted with trolling, where baited fishing lines instead of trawls are drawn through the water. Trolling is used both for recreational and commercial fishing whereas trawling is used mainly for commercial fishing. Trawling is also commonly used as a scientific sampling, or survey, method.

BOTTOM VERSUS MIDWATER TRAWLING

Trawling can be divided into bottom trawling and midwater trawling, depending on how high the trawl (net) is in the water column. Bottom trawling is towing the trawl along (benthic trawling) or close to (demersal trawling) the sea floor. Midwater trawling is towing the trawl through free water above the bottom of the ocean or benthic zone.

Midwater trawling is also known as pelagic trawling. Midwater trawling catches pelagic fish such as anchovies, shrimp, tuna and mackerel, whereas bottom trawling targets both bottom living fish (groundfish) and semi-pelagic fish such as cod, squid, halibut and rockfish.

The gear itself can vary a great deal. Pelagic trawls are typically much larger than bottom trawls, with very large mesh openings in the net, little or no ground gear, and little or no chaffing gear. Additionally, pelagic trawl doors have different shapes than bottom trawl doors, although doors that can be used with both nets do exist.

NET STRUCTURE

When two boats are used (pair trawling), the horizontal spread of the net is provided by the boats, with one or in the case of pelagic trawling two warps attached to each boat. However, single-boat trawling is more common. Here, the horizontal spread of the net is provided by trawl doors (also known as "otter boards"). Trawl doors are available in various sizes and shapes and may be specialised to keep in contact with the sea bottom (bottom trawling) or to remain elevated in the water. In all cases, doors essentially act as wings, using a hydrodynamic shape to provide horizontal spread. As with all wings, the towing vessel must go at a certain speed for the doors to remain standing and functional. This speed varies, but is generally in the range of 2. 5–4. 0 knots.

The vertical opening of a trawl net is created using flotation on the upper edge ("floatline") and weight on the lower edge ("footrope") of the net mouth. The configuration of the footrope varies based on the expected bottom shape. The more uneven the bottom, the more robust the footrope configuration must be to prevent net damage. This is used to catch shrimp, shell fish, cod, scallops and many others. Trawls are funnel shaped nets that have a closed off tail where the fish are collected and is open on the top end as the mouth. Trawl nets can also be modified, such as changing mesh size, to help with marine research of ocean bottoms.

ENVIRONMENTAL EFFECTS

Although trawling today is heavily regulated in some nations, it remains the target of many protests by environmentalists. Environmental concerns related to trawling refer to two areas: the lack of selectivity and the physical damage which the trawl does to the seabed.

SELECTIVITY

Since the practice of trawling started (around the 15th century), there have been concerns over trawling's lack of selectivity. Trawls may be non-selective, sweeping up both marketable and undesirable fish and fish of both legal and illegal size. Any part of the catch which cannot be used is considered by-catch, some of which is killed accidentally by the trawling process. By-catch commonly includes valued species such as dolphins, sea turtles, and sharks, and may also include sublegal or immature individuals of the targeted species.

Many studies have documented large volumes of by-catch that are discarded. For example, researchers conducting a three-year study in the Clarence River found that an estimated 177 tons of by-catch (including 77 different species) were discarded each year.

Size selectivity is controlled by the mesh size of the "cod-end"—the part of the trawl where fish are retained. Fishermen complain that mesh sizes which allow undersized fish to escape also allows some legally–catchable fish to escape as well.

There are a number of "fixes", such as tying a rope around the "cod-end" to prevent the mesh from opening fully, which have been developed to work around technical regulation of size selectivity. One problem is when the mesh gets pulled into narrow diamond shapes (rhombuses) instead of squares. The capture of undesirable species is a recognised problem with all fishing methods and unites environmentalists, who do not want to see fish killed needlessly, and fishermen, who do not want to waste their time sorting marketable fish from their catch. A number of methods to minimize this have been developed for use in trawling. Bycatch reduction grids or square mesh panels of net can be fitted to parts of the trawl, allowing certain species to escape while retaining others. Studies have suggested that shrimp trawling is responsible for the highest rate of by-catch.

ENVIRONMENTAL DAMAGE

Trawling is controversial because of its environmental impacts. Because bottom trawling involves towing heavy fishing gear over the seabed, it can cause large-scale destruction on the ocean bottom, including coral shattering, damage to habitats and removal of seaweed.

The primary sources of impact are the doors, which can weigh several tonnes and create furrows if dragged along the bottom, and the footrope configuration, which usually remains in contact with the bottom across the entire lower edge of the net. Depending on the configuration, the footrope may turn over large rocks or boulders, possibly dragging them along with the net, disturb or damage sessile organisms or rework and re-suspend bottom sediments. These impacts result in decreases in species diversity and ecological changes towards more opportunistic organisms. The destruction has been likened to clear-cutting in forests.

The primary dispute over trawling concerns the magnitude and duration of these impacts. Opponents argue that they are widespread, intense and long-lasting. Defenders maintain that impact is mostly limited and of low intensity compared to natural events. However, most areas with significant natural sea bottom disturbance events are in relatively shallow water. In mid to deep waters, bottoms trawlers are the only significant area-wide events.

Bottom trawling on soft bottoms also stirs up bottom sediments and loading suspended solids into the water column. One bottom trawler can put more than 10 times the amount of suspended solids pollution per hour into the water column than all the suspended solids pollution from all the sewerage, industrial, river and dredge disposal operations in Southern California combined.

These turbidity plumes can be seen on Google Earth in areas where they have high resolution offshore photos. When the turbidity plumes from bottom trawlers are below a thermocline, the surface may not be impacted, but less visible impacts can still occur, such as persistent organic pollutant transfer into the pelagic food chain.

As a result of these processes, a vast array of species are threatened around the world. In particular, trawling can directly kill coral reefs by breaking them up and burying them in sediments. In addition, trawling can kill corals indirectly by wounding coral tissue, leaving the reefs vulnerable to infection. The net effect of fishing practices on global coral reef populations is suggested by many scientists to be alarmingly high. Published research has shown that benthic trawling destroys the cold-water coral *Lophelia pertusa*, an important habitat for many deep-sea organisms.

Midwater (pelagic) trawling is a much "cleaner" method of fishing, in that the catch usually consists of just one species and does not physically damage the sea bottom. However, environmental groups have raised concerns that this fishing practice may be responsible for significant volumes of by-catch, particularly cetaceans (dolphins, porpoises, and whales).

REGULATION

In light of the environmental concerns surrounding trawling, many governments have debated policies that would regulate the practice.

OTHER USES OF THE WORD "TRAWL"

The noun "trawl" has many possibly confusing meanings in commercial fisheries. For example, two or more lobster pots that are fished together may be referred to as a trawl. In some older usages "trawling" meant "long-line fishing"; that usage occurs in Rudyard Kipling's book *Captains Courageous*. (This use is perhaps confused with trolling, where a baited line is trailed behind a boat. Troll also has several meanings.)

The word "trawling" has come to be used in a number of non-fishing contexts, usually meaning indiscriminate collection with the intent of picking out the useful bits. For instance, in law enforcement it may refer to collecting large volumes of telephone call records hoping to find calls made by suspects. It also occurs frequently in reference to research methods, where it means searching through written sources for relevant information.

8

Fish Capture Technology

ECONOMICS OF FISHING OPERATIONS

Revenues and costs mainly determine the economics of fishing operations. Revenues depend on species and quantities caught and prices obtained, which again depend on marketing channels and markets, seasonal fluctuations and other factors. The main cost factors are capital investment and operation costs, which can be divided in labour costs, running costs and vessel costs. The major components of labour costs are wages and other labour charges such as insurance and employer' s contributions to pensions funds. Running costs are principally composed of fuel, lubricants, cost of selling fish, harbour dues, cost of ice, food and supplies for the crew. The major elements of vessel costs are vessel and gear repair and maintenance expenses and vessel insurance.

In addition, fishing operations also carry external costs which are sometimes difficult to quantify. External costs are defined as costs which are created by a fishing enterprise for others, *i. e.* other enterprises or society, for example through depletion of fish stocks or destruction of the coastal ecosystem. The economic and financial performance of fishing operations is generally assessed with the help of two indicators. For the assessment of the economic performance of a fishing vessel, as in the case of other economic enterprises, the ratio between net cash flow and total earnings (NCF/TE) is used. This ratio is a general indicator of economic profitability/viability of enterprises as it shows the amount of total earnings required by a certain type of fishing vessel in order to generate a given amount of net profit. The financial performance is assessed with the help of the rate of return on investment (ROI). The ratio shows how much money needs to be invested in a fishing enterprise in order to generate a certain net profit.

The economic development and performance of fishing operations is affected by various factors including fluctuations in revenue, perishability of product, falling yields *i. e.* catch per unit of effort, static or falling demand, unforeseen increases in the cost of key inputs and catch and effort restrictions. Theory suggests that in an open-access, unregulated fishery, the fishery will eventually end up producing at the point where total revenue equals total costs. 1999 and 2000, FAO conducted 15 country level studies on the economic and financial performance of marine capture fisheries in close co-operation with fisheries research institutions and national fisheries administrations in selected countries in Asia, Africa, Latin America and Europe.

The findings suggest that, despite fully and sometimes over-exploited fisheries resources, in most cases marine capture fisheries is an economically and financially viable undertaking, which generates sufficient revenue to cover the cost of depreciation as well as the opportunity cost of capital to generate funds for reinvestment in addition to generating employment, income and foreign exchange earnings. Of the 108 types of fishing vessels studied in 15 South American/Caribbean, European, African and Asian countries, 105 or 97 % had a positive gross cash flow and fully recovered their cost of operation. Only three types of vessels *i. e.* stow-nettersinChina and and semi- industrial and industrial shrimp and bottom fish trawlers in Trinidad and Tobagoshowed operational losses. When also considering the cost of capital *i. e.* the cost of depreciation and interest, 92 out of the 108 types of vessels or 85 % showed a net profit after deducting the cost of depreciation and interest.

In the 10 countries, which participated in the previous as well as in the recent study, two countries showed marked improvements in the profitability of their fishing vessels *i. e.* France and Spain while two countries showed a declining profitability *i. e.* the Peoples Republic of China and Germany. In the remaining six countries *i. e.* the Republic of Korea, Indonesia, India, Senegal, Argentina and Peru, the situation was similar as during the previous study carried out between 1995 and 1997.

These overall positive results were also achieved because of higher prices paid to producers as compared to the previous study period. There were only few indications that fishing effort had been reduced and fish stocks had recovered. It was also observed that some fishing fleets had adapted themselves to new conditions dictated by depleted and changing abundance of resources and new access to markets in the context of globalisation by changing their fishing operations.

Those vessels, which had previously shown positive results but now incurred losses, were generally older vessels due to the fact that they continued to work on overexploited stocks. Regarding the impact of government financial transfers it was found that in two countries, belonging to the European Union and in India, almost all types of vessels covered by the cost and earnings study, which received subsidies, would also have been profitable without subsidies. In the Republic of Korea, the situation was mixed while in Thailand, those types of vessels. which could avail themselves of tax exemptions on fuel needed the tax exemption in order to have a positive gross cash flow. No subsidies were available in Indonesia.

In the case of other countries it was observed that no detailed empirical information was available neither on the amount of government financial transfers to the fishing industry nor on the financial performance of individual fishing enterprises. There is a need to collect and analyse information on the quantity and impact of government financial transfers on the economic performance of fishery enterprises with a view to reduce and rationalise the use of scarce government financial resources and to avoid that subsidies have a negative impact on fisheries resources, on the coastal environment and on trade.

Electronic Instruments

The command console on a larger fishing vessel is at the center of the bridge, with a clear view all around. The compass or a repeater is well- displayed for the benefit of the helmsman. The autopilot is built into the steering console with all the associated controls (*i. e.* rudder position, rate of turn indicator). On a fishing vessel the equipment is normally clustered around the skipper's chair but in the largest vessels an additional chair is provided for a co-pilot. In fact, the layout is remarkably similar to the cockpit design of a modern aircraft, with the instruments surrounding the pilot and co-pilot. The main displays are shown on monitors in front of the control position and increasingly these are shown on one large integrated display. Instruments or monitors that are used more infrequently are mounted in the deckhead (*i. e.* ceiling). Probably the only differences from the layout of an aircraft cockpit is that there is more space available on the bridge of a fishing vessel and the bridge windows are tilted downwards rather than upwards.

The instruments spread around the skipper can be grouped into several relatively-loose categories. The first is navigational instruments which are used for the navigation of the vessel while at sea and in harbour. Other instruments are used for fish detection and during the fishing operation, including sonar equipment and electronic aids for the operation. Radio communications are vitally important for the safety and for general communications, and as such are situated around the control position. Communications within the vessel are also provided at this point with links to the messroom, the skipper's cabin, engine room and key points on the working deck of the vessel.

Tension meters built into the winches allow the skipper to monitor the fishing gear and display information, which indicates any malfunction. Trawlershave the tension on both towing warps displayed so that if the gear comes fast on the bottom an immediate alarm is sounded. Differing tensions on the two warps indicate a malfunction unless this is caused by a change in course. In fact, modern computerized winch controls allow for changes in the warp length, to ensure that the trawl remains in a fishing mode throughout a change of course. In purse seines, sensors give the rate of sinking of different parts of the net during setting which can be important. Other types of fishing gear such as gill netsand longlinescan be installed with automatic readouts fitted on the reels and drums which indicate the length of fishing gear which has been set.

CAPTURE FISHERIES

Capture fisheries is intended for catching fishes and also prawns, lobsters, crabs, sea-cucumbers, shales, pearl oysters, edible bivalve and copious other organisms of other than fishes etc. Premitive human being were acquinted with capture fishery centuries passed for him to observe and understand for the possibilities of culturing fish. Then also he depended mostly on the culture of fishes with parental care.

Later, he tried to collect the fingerlings in canals, distribution canals. In the earlier days, the mixture of carnivore fish fingerlings and carp fish fingerlings were stocked together in tanks. Later, they were segregated and stocked selecting the required vaiety. Capture of fishes can be broadly divided in to two types; a) Capture by Human effort b) Capture by observing the behavioural pattern of Fishes.

Inland capture fisheries of India has an important place; it contributes to about 30% of the total fish production. The large network of inland water masses will continue to provide great potential, for economic capture fishery which consequently will compete well with fast/growing fish-culture practices. The freshwater inland water bodies fall into five major categories, distinguished as the Ganga, the Brahmaputra and the Indus system of the Northern India, and the East and the West coast river systems of the Southern (peninsular) India. These river systems have certain characteristics of their own with respect to their ecology, climatic conditions and fish populations of commercial food fishes. Besides, there are a number of land-locked lakes especially those situated at high altitudes which have started supporting cold water fisheries of both indigenous and exotic species.

In addition to the above-mentioned freshwaters, there are also rich fisheries offered by extensive brackishwaters, including important estuaries (Hooghly-Matlah, Mahanadi and Godavari estuaries), lagoons (Chilka lake, Pulicat lake) and backwaters (Vembanad) and paddy fields (Pokkali in Kerala). Chilka lake in the state of Orissa is an open shallow brackishwater lake having an area of 906 sq. km. in summer and 1165 sq. km. in rainy season. A long canal joins it with sea. Waters from river Daya (Mahanadi) and other smaller streams flow into it. Recent additions to the natural inland water bodies are man-made reservoirs. There are at present some 300 reservoirs which hold very good prospects, after restocking, both for capture as well as for culture fisheries. Some of these reservoirs have responded fairly well to attempts to restock them with indigenous as well as exotic species.

Inland capture fisheries is a-continually expanding industry bringing under its fold newer fisheries of a local or regional nature, while improving upon those which are existing already. Introduction of exotic species from abroad and inter-regional transplantation of fish from Northern to Southern waters have been most welcome and rewarding.

INLAND CAPTURE FISHERY

The inland capture fishery, however, stands at a critical juncture, which draws a special attention at the national level. Rapid industrialization movements in the country have given a serious blow to the growth of the inland fisheries which was struggling to come out of the old-fashioned style to a more rational and scientific style. Construction of dams have been the cause of decline and damage to several regionally important fisheries. Discharge from industrial establishments, multiplying at mushroom growth, into inland water bodies is polluting the water in very serious proportions, and is damaging the fish populations tremendously.

Already, old-age practices of indiscriminate fishing of fingerlings and juveniles, supporting local and seasonal fisheries, especially in breeding or nursery gro. unds, have been doing enormous damage, and needed effective controls for conservation. Likewise, time-old practice of sewage disposal into rivers was a menacing practice causing heavy pollution. Great harm is also being donc from agricultural wash coming to inland waters, which brings to fish a very toxic principle of the numerous pesticides used in the agricultural practices. Food even more than clothing or shelter is the indispensable necessity of mankind. The history of our species is largely an account of the evolution of equipment and methods of hunting, gathering, cultivating, breeding and otherwise controlling food supplies. Although the initial production of food is primary, the methods adopted for its preservation, storage and distribution, are no less important and react on the former.

In all periods and at all levels of technological progress, fish has usually played an essential part in man's diet. Although there are occasionally communities that do not eat fish, whether for reasons of geography far more people throughout the world have been and still are dependent almost entirely on fish than is the case with meat. Fresh fish flesh provides an excellent source of protein for human diet. This protein is relatively of high digestibility, biological and growth promoting value for human consumption. Nutritional studies have proved that fish proteins rank in the same class as chicken protein and are superior to milk, beef protein and egg albumen. Fish proteins comprise all the ten essential amino acids in desirable strength for human consumption. This accounts for the high biological value of fish flesh.

Fish flesh therefore becomes a valuable supplement to human diet for people who are habitually taking cereals, starchy roots and sugar as their principal diet. Besides proteins, fish flesh also offers minerals. Iodine, vitamins and fat, over and above all, fish flesh cooks easily, offers a palatable taste and flavour, and is easily digestible. Fish is consumed either as a preparation from freshly caught fish or from those that have been preserved in some farm. Fishes are consumed as food in fresh condition. Some of them are also utilized after the preservation. Fish, however, is more susceptible to spoilage than certain other animal protein foods, such as meat and eggs. As part of the natural process by which organic matter is broken down and returned to the nitrogen cycle, fish flesh is rapidly invaded, digested and spoiled by the micro-organisms which are abundant on the skin and in the intestines. Ferments ('enzymes' to the scientist) also contribute to the dissolution, and oxidation by atmospheric oxygen is an additional process of deterioration, particularly in the case of natural fats.

PREVENT SPOILAGE OF FISH

To prevent spoilage of fish, some form of preservation is necessary. Preservation means keeping the fish, after it has landed, in a condition wholesome and fit for human consumption for a short period of a few days or for longer periods of over a few months. During the period of preservation the fish is kept as 'fresh' as possible, with minimum losses in flavour, taste, odour, form, nutritive value, weight and digestibility of flesh. This preservation should cover the entire period from the time of capture of fish to its sale at the retailer's counter.

As a result, methods of preservation to counteract these processes must in former days have been essential to the utilization of fish as food. Fish in early times occupied a key position as one of the most easily accessible sources of protein food, and the spread of man himself was probably determined by the success of the techniques of preservation and storage employed. There are regions in the world where fish preservation is not necessary. In arctic zones, for instance, landed fish is rapidly frozen due to extremely cold climate. No further preservation is called for. In temperate climate also, the need for preservation is not so pressing because fish can remain "fresh" for a few days without any preservation. The picture is different for tropical areas of the world. Here the hot climate favours rapid spoilage of fish.

The principal processes employed to check bacterial and other forms of spoilage have been few in number. 'Curing' by drying, smoking, and salting with common salt, in various combinations, is bacteriostatic or bactericidal in varying degree. Other salts, acids such as vinegar, sugar, and certain spices and herbs, have a similar effect. Even fermentation itself can be a means of preservation if the process of decomposition is properly controlled. Cooling by means of ice slows down the multiplication of micro-organisms and hard-freezing below a certain temperature suspends it altogether. Canning results in their destruction by heat. These processes, which are all in use to-day have their characteristic problems and limitations. Furthermore, a certain level of technique, and of social organization, was required in each case before material circumstances permitted its appearance in history.

During preservation and processing, some materials of fish and prawn are discarded as waste. Similarly some trash and distasteful fishes are unsuitable for human consumption. These waste material and above fishes become an important source to produce fish by-products. Which in turn are used to produce different useful fish by-products by different fish by-product industries. After fish harvesting the fishes are sold in the market freshly or after the preservation. In earlier days, the term marketing of fish meant buying and selling of fish at the landing centre. After the second world war, the concept and function of fish marketing have taken a new role in business activities. The fisheries have now become highly industrialized in all advanced fishing nations. The new marketing techniques have been adopted so as to sell more fish. The modern fish marketing system lays emphasis in meeting the existing demands for fish, besides tapping the potential demand in the important markets. In many advanced countries the improved methods of fish marketing are being adopted with the advancement of fisheries development. A progressive fish marketing system will also provide remunerative price to the primary producer though the interest of the consumer is also protected.

In many developing countries traditional system and fish marketing is adopted. The methods and practices in trade dealings are based on some customs. These practices have remained unchanged and unimproved over decades. The fish marketing is normally done at the collection centres, which are mainly situated in the area of fish landing. Fish has peculiar feature at its own and gives a big strain and stress on the method of its marketing. The fish marketing should not have the object of only catching and selling of fish but should have the wide scope for exploitation, production, distribution, preservation and transportation of fish in addition to actual sale of fish by reducing middlemen.

MAKING OF CAPTURE FISHERIES

UNCED and its outcomes are said to be a new beginning for international lawmaking, marking the transition from international environment law and international economic law to an international law of sustainable development. In respect of fisheries law, particularly marine capture fisheries, calls on states "to pursue the protection and sustainable development of the marine and coastal environment and its resources" in accordance with the 1982 UN Convention of the Law of the Sea.

After UNCED, the international community adopted a two-tiered approach to deal with the problems of over-fishing. The first was the negotiation of international agreements on specific marine fish stocks and high seas fishing, while the second, which had commenced prior to UNCED, was the development of soft law instruments for the conservation and management of fisheries.

An intergovernmental conference under the auspices of the United Nations to promote effective implementation of the 1982 UN Convention and to take effective action to deter reflagging of fishing boats. Pursuant to this commitment, the UN Conference on Straddling Fish Stocks and Highly Migratory Fish Stocks in 1995 adopted the Agreement for the Implementation of the Provisions of the United Nations Convention of the Law of the Sea of 10 December 1982 Relating to the Conservation and Management of Straddling Fish Stocks and Highly Migratory Fish Stocks (UN Fish Stocks Agreement). At the same time, FAO was working on an agreement to reduce fishing on the high seas contrary to internationally agreed conservation and management measures. In November 1993, the FAO Conference adopted the Agreement to Promote Compliance with International Conservation and Management Measures by Fishing Vessels on the High Seas (Compliance Agreement).

In the meantime, a work plan had been developed in 1992 for the elaboration of a Code of Conduct for Responsible Fisheries. In the run-up to UNCED, a Conference on Responsible Fishing, held in Cancún, Mexico in May 1992, adopted the Cancún Declaration on Responsible Fishing, which called upon FAO to begin development of such a Code. The final text of the Code was adopted in November 1995. The Compliance Agreement was intended to be an integral part of the Code.

MARINE FISHERIES AND HIGH SEAS FISHING

The Code of Conduct is intended to cover much more than marine fisheries and high seas fishing. FAO has produced non-legal technical guidelines to provide general advice in support of the implementation of the Code. These include guidelines on fishing operations; vessel monitoring systems; the precautionary approach to capture fisheries and species introduction; the integration of fisheries into coastal area management; fisheries management; conservation and management of sharks; aquaculture development; good aquaculture feed manufacturing practice; inland fisheries; responsible fish utilization; and indicators for sustainable development of marine capture fisheries.

The Code of Conduct was followed by four international plans of action developed under the auspices of FAO: the International Plan of Action for Reducing Incidental Catch of Seabirds in Longline Fisheries, the International Plan of Action for the Conservation and Management of Sharks, the International Plan of Action for the Management of Fishing Capacity and the International Plan of Action to Prevent, Deter and Eliminate Illegal, Unreported and Unregulated Fishing. Other important international fisheries instruments are the Rome Consensus on World Fisheries, the Rome Declaration on the

Implementation of the Code of Conduct for Responsible Fisheries and the Kyoto Declaration and Plan of Action.

The influence of these international law developments and instruments on national legislation cannot be underestimated. Not only have they given impetus at the national level for practical implementation of environmental concepts such as sustainable development and the precautionary principle or precautionary approach in the field of fisheries, but states have also acted to give effect to specific provisions of international fisheries instruments, particularly the Compliance Agreement and the UN Fish Stocks Agreement. Both Agreements seek to enhance responsible fishing or conservation and sustainable use of living marine resources although they emphasize different means to that end.

The Compliance Agreement focuses on ensuring compliance with international conservation and management measures through effective use of flag state responsibility (*i. e.* , ensuring that vessels do not fish on the high seas without authorization). The UN Fish Stocks Agreement reinforces flag state responsibility but emphasizes that conservation and management of straddling and highly migratory fish stocks shall be undertaken through cooperation facilitated largely by regional fisheries bodies (RFBs). This latter requirement, along with the further restriction that non-members of a fisheries management body shall not have access to the fishery resources to which the measures of that fishery management body apply, has had an extraordinary impact in fisheries conservation and management in restricting freedom of fishing on the high seas.

The hard law and soft law instruments adopted in recent years influence and reinforce one another. The preambles of the two Agreements refer to Agenda 21 and to the issues identified in the Cancún Declaration. They also use the term "responsible fishing", which is clearly drawn from the Code of Conduct. Equally, soft law instruments such as the Code of Conduct call on states to become parties to fisheries agreements and to implement them. These connections give a basis for using soft law instruments as aids to interpretation.

FISHERIES TECHNOLOGY

Harvesting of wild aquatic resources and production in controlled environments (aquaculture) are done through the use of various technologies - from artisanal to highly-industrial - encompassing vessels and equipment, fishing gears and their operation and different types of enclosures to raise fish and other aquatic products.

For both capture fisheries and aquaculture, the technological development and widespread use of synthetic fibres, hydraulic equipment for gear and fish handling, electronics for fish finding, satellite-based technology for navigation and communications, onboard conservation and increased use of outboard engines in small-scale fisheries have all contributed to the major expansion of fisheries and aquaculture in recent decades. Technical advances have generally led to more efficient and economical fishing operations, reduction of the physical labour required per unit of output and improved access to resources.

Where management has been ineffective, the greater efficiency of fishing methods and aquaculture production has sometimes led to overfishing and environmental degradation. This points to the need to develop more effective fisheries management frameworks, together with safer and more environmentally-friendly methods of production, for example, in developing selective fishing gear and in designing aquaculture systems that reduce their impact on external environments.

FISHING VESSELS

The vast majority of commercial vessels load and discharge cargo in the safety of ports; their main function at sea is transportation. A fishing vessel differs in that it is used to hunt, locate, catch, load (and sometimes discharge), as well as process and conserve cargo at sea, all in variable weather conditions. In effect, it is a place of work and is a very specialized vessel which is intended to perform all these well defined tasks. The size, deck layout, carrying capacity, accommodation, machinery and equipment of fishing vessels are all related to its function in carrying out its planned operations.

Factors which influence the design of a fishing vessel may be grouped under the following headings:

- The species, location, abundance and dispersion of the fish resources
- Fishing gear and methods
- Geographical and climatic characteristics of the fishing area
- Seaworthiness of the vessel and safety of the crew
- Handling, processing and stowage of catch
- Availability of finance
- Availability of boatbuilding and fishing skills
- Laws and regulations applicable to fishing vessel design, construction and equipment
- Choice and availability of construction materials
- Economic viability

Because of the inherent variations in each of these factors, the diversity of fishing vessels designs operating around the world is enormous, ranging from 2 metre dug out canoes to factory trawlers exceeding 130 metres in length, with trip durations ranging from a few hours to over a year.

CAPTURE FISHERY RESOURCES IN INDIA

India is endowed with vast and varied aquatic resources (marine and Inland) amenable for capture fisheries and aquaculture. While the marine water bodies are used mainly for capture fisheries resources, the inland water bodies are widely used both for culture and capture fisheries. Most of the inland water bodies are captive ecosystems where intensive human intervention in the biological production process can be possible and thereby holding enormous potential for many fold increase in fish output. Inland water bodies include freshwater bodies like rivers, canals, streams, lakes, flood plain wetlands or beels (ox-bow lakes, back swamps, etc.), reservoirs, ponds, tanks and other derelict water bodies, and brackish water areas like estuaries and associated coastal ponds, lagoons (Chilka lake, Pulicat lake) and backwaters (vembanad backwaters), wetlands (bheries), mangrove swamps, etc.

Of these, the rivers, canals, streams, lakes, large and medium reservoirs, estuaries, and associated backwaters and lagoons support the capture fisheries. Whereas freshwater ponds, tanks, swamps and estuarine wetlands (bheries), paddy fields, small shallow coastal lagoons and coastal pond farms support the culture fisheries or aquaculture.

In capture fisheries, the wild populations are simply harvested from the natural waters with little human intervention in modifying the ecosystem *i. e.* hunting. Example: marine fishery. On the other hand, in a culture fishery, the whole operation is based on captive stocks with a high degree of effective human control over the water quality and other habitat variables. Example: Culture of fish and shell-fish in ponds. When the fish harvest in an open water system depends solely or mainly on artificial recruitment (stocking), it is generally referred to as culturebased fisheries. Culture-based fishery is the most common method of enhancing the fish production being followed in some inland water bodies in India.

RIVERINE FISHERIES

India is blessed with a vast inland water resources in the form of rivers, estuaries, natural and man made lakes. The Inland water bodies have been divided into five riverine systems and their tributaries extending to a length of about 29,000 km in the country – Indus, Ganges, Bramhaputra, East flowing riverine system and West riverine system. All these rivers, their tributaries, canals and irrigation channels have and area of roughly 13000km. These water bodies harbour the original germplasm of one of the richest and diversified fish fauna of the world comprising 930 fish species belonging to 326 genera. The major river systems of India on the basis of drainage can be divided broadly into two major rivers systems. They are (i) Himalayan rivers system (Ganga, Indus and Bramhaputra) and (ii) Peninsular river system (East cost and West coast river system).

Ganges River System

It is the largest river systems of the world, having a combined length (including tributaries) of 12,500 km. It originates from Gangotri in the Himalayas at a height of about 3129 km above the sea level. After origin it drains the southern slopes of the central Himalayas. Ganga passes through UP, Bihar, some parts of Rajasthan, M. P. and west Bengal and finally joins to the Bay of Bengal. It has a large number of tributaries and 'Yamuna' river is one of the major tributaries of this system, which is about 1000 km long. The other tributaries are – Ram Ganga. Gomti, Ghaghra, Gandak, Kosi, Chambal, Betwa and Ken. Further more; it has numerous lakes, ponds and Jheels, both perennial and seasonal areas. It has a total catchment area of 9. 71 lakh sq. km and receives an annual rainfall of 25-77 inches.

Physico-Chemical Characteristics:

- Temperature range-16. 70C in January – 31. 50C in June to sept.
- PH-7. 4 during June to August and Maximum 8. 3 during January to May.
- Turbidity-100 ppm in January; 1100-2170 ppm during July to September.
- Do2-5. 0 to 10. 5 ppm during January to February while in monsoon 2. 00ppm (July-Sept.)
- Co2-0. 6 ppm-10. 0ppm
- Chloride-4. 0-35. 4 ppm
- Phosphate-0. 05-021ppm
- Nitrates-0. 08-0. 22ppm
- Silicates-4. 0-20. 3ppm
- Carbonates-1. 0 – 12. 0 ppm

Common Phytoplanktons: Phytoplanktons are generally poor during the monsoon and autmn months. Common phytoplanktons found in Ganga river system are – (i) Members of Bacillariophyceae like Amphora, Asterionella, Cymbella, Navicula and Synedra etc. (ii) members of Chlorophycace like, Chlorella, Closterium, Denticula, Pandorina and Spirogyea etc. (ii) members of Myxophyceae like Anabaena, Nostoc, Oscillatoria etc.

Common zooplanktons: Rattulas, Rotaria, Keratella, Filuia, Notops, Monostyla etc.

Fisheries of Ganga river systems: The Ganga river system supports a large number of commercially important fish species including major carps (*Labeo. rohita: L. Calabasu, Catla catla and Cirrhinus mrigala*), minor carps (*Labeo fimbriatus; L. bata; Cirrhinus. reba*), catfishes (*Wallago. attu; Mystus. aor; M. . tengara, Clarias. batrachus; Heteropneustes fossilis*), cluipeiods, murrels (*Channa* species), feather backs (*Notopterus. notopterus; N. chitala*), mullets (*Mugil corsula*), fresh water eel (*Anguilla*) and prawns (*Macrobrachium malcolmsonii; Palaemon. Lamarii*). Apart from these fishes, the others like *Pangasius*; *silonia silondia*; *Gudusia chapra; Bagasius. bagasius; Eutropichthys. vacha* are also found in the river system.

The commercial fisheries in this zone are non-existing due to spares population, inaccessible terrain and poor communication between fishing grounds and landing centers. The fish yield has been declined over the years due to 1) sandification of the river bed (upto Patna) which reduced the rivers productivity due to blanket effect, (2) marked reduction in the water volume on account of increase sedimentation, (3) increased water abstraction and (4) irrational fishing. In spite of this, the Ganga river system is contributing nearly about 89. 5% of the total fish seed correlation of India.

Fishing gears used: The principal gears used in Ganga river system are dragnets, cast nets and bag nets.

Godavari River System

It originates in Doolai hills near Nasik in North Western Ghats. This river system is a part of East coast of pennensular river system, with a length of 1465 km covering the states like Maharastra, Andhrapradesh and Madhyapradesh. It has the primary tributaries like manjira, Wainganga; Subtributaries like paingunga and wardha and minor tributaries like maner and sabari. It drains into Bay of Bengal. It has a total catchments area of over 315,980 sqkm.

Physico-Chemical characteristics:

- Temperature – 27. 5 to 36. 40C
- PH-7. 2 to 8. 3
- Do2 mg/L-1. 26-18. 2
- Co2-0. 0 – 6. 6 ppm
- Bicarbonates-45. 8-192-2ppm

Fisheries of Godavari River System: The head waters harbour a variety of game fishes but don't support the commercial fisheries. The commercial fisheries consist of carps (major caps, Labeo fimbriatus), large cat fish (Mystus spp. , Wallago attu; Bagarius bagarius) and fresh water prawn (Macrobrachium rosenbergii). Hilsa formed lucrative –fisheries and the Indian major carps planted in the river in the beginning in 19th century are thriving well and contributing to the commercial fisheries.

Fishing gears used: The principle gear used in Godavari river system are falls under two categories *viz.* , gill nets, which include setgill nets, drift nets, drag gill nets (Benduvala) and the barrier gillnet (Katu vala). Seines include shore seine (Jaruguvala), Large seine (Allui vala) and dragnet castnets are also employed for fishing.

Krishna river System

The originates in Western Ghats region, south of Poona and finally drains into East coast, with an a length of 1401km covering the state like Maharastra, Karnataka and Andhrapradesh. It has the main tributaries like Bhima (Annual) and Tungabhadra (Perennial). This river system has an total catchments area of 2,33,229 sq km. The physico-chemical characteristics, fish fauna and the fishing gears used similar to the Godavari river system. In general, the physiographic and fish fauna resembles the Godavari river systems. The head waters support rich fishery when compared to mid-stretch, which is rocky and inaccessible. No information is available on its present fishery and catch statistics.

Cauveri River System

This river system originates from Brahmagiri hills on western ghat, with an elevation of 1340 m extending to a length of 800 km. this river system covering the states the Karnataka and Tamil Nadu finally drains into Bay of Bengal in Thanjavur district of Tamil Nadu. It has the tributaries like Bhavani, Noyil and Amaravathi. This river system has an total catchment area of 4,70000 sq km. The water resources of the river are extensively exploited as numerous reservoirs, anicuts and barrages have been built on the river.

Physico-Chemical characteristics:

- Temperature – 26 to 30. 90C
- PH-7. 6 to 8. 5
- Do2 mg/L-1. 26-18. 2
- Co2-0. 0 – 6. 6 ppm
- Bicarbonates-45. 8-192-2ppm

Fisheries of Cauveri river system: The Cauveri river system exhibits substantial variations in its fauna-nearly 80 species of fish belong to 23 families have been reported from this river system. Its fish funa differs significantly from Godavari and Krishna river system. The fishes like *Acrossocheilus hexagonolepis; Tor. Putitora; Barbus carnatus*; *B. dubius; Labeo kontius; L. ariza;* Cirrhinus cirhosa; Mystus aor; Mystus seenghala; Pangasius pangasius; *Wallago attu; Silonia silonida; Glyptothorax madrapatanus;* Gangetic carps such as *Catla catla; Labeo rohita; Cirrhinus mrigala and the exotic species Cyprinus carpio and Osphronemus goramy* have been transplanted in Cauveri river system. The game fish *Tor khudri* and *T mussullah* are found all along the river length except the deltaic stretch.

West Coast River System

The west coast system comprises the river Narmada and Tapti, both of which flow in westernly direction of the country and drain the narrow belt of peninsular India, west of the western ghats further in the north the system forms basins of Narmada and Tapti and the drainage of Gujarat.

Narmada river system: This river system originates in Amarkantak hills of Madhya Pradesh, at an elevation of 1,057 m above the sea level. The length of the river is 1280 km, covering Madhyapradesh and Gujarat states and finally drains into gulf of Cambay in Gujarat. The effective catchments area of this river system is 94235 sq. km and 6330 sq. km of its all tributaries. This river system comprises of total 18 tributaries, of which 16 in Madhya Pradesh and 2 in Gujarat. This river system receives and annual rain fall of 12" – 115".

Fishery of Narmada river system: Narmada river harbours 84 fish species belonging to 23 genera. The contribution of carps in commercial fishery is of the order of 60. 4%, followed by catfishes of 34. 1 % and miscellaneous fishes of 5. 5%. The carp fish groups are *Tor tor; Labeo. frimbriatus; L. calabasu; L. bata; L. gonius; Cirrhinus. reba; Puntius. sarana* etc, cat fish groups such as Mystus seenghala; M. aor; M. cavasius; Wallago. Attu; Clupisoma *garua; Ompak bimaculatus* and miscellaneous fish groups like *Channa* spp, *Mastacembalus* spp; *Notopterus notopterus.* Cast nets, gill nets and longlines are the fishing gears that are generally used in these waters.

Tapti River System: This river system originates in Mount Vindhya of Satpura range at an elevation of 670 –100m above the sea level, with a total length of 720 km. This river system covers the states like Maharastra, Madhya Pradesh and Gujarat and finally drains into Arabian Sea at Dumas near Surat in Gujarat. The total catchments area of this river is 48,000 sq. km and annual rainfall is more or less similar to the Narmada river system. Not much information of fish stock composition and fish yield is available. The main fisheries of this river system are *Tor. tor; Mystus.* seenghala; M. aor; Wallago attu; Labeo calabasu; L. fimbriatus; Puntius *sarana; Cirrhinus mrigala; C. reba, Chupisorna garna; Channa* spp; Mastacembalus. armatus. Cast nets, gillnets and long lines and also Mahajal is used as the fishing gear in these waters.

Factors Influencing Fish Yield from Rivers

The intensity of fishing, nature of exploitation and species orientation are the characteristics of the artisan riverine fisheries and are governed by

- Seasonality of riverine fishing activity
- Unstable catch composition
- Conflicting multiple use of river water
- Cultural stresses leading to nutrient loading and pollution.
- Lack of understanding of the fluvial system and infirm data base. vi) Fragmentary and out molded conservation measures lacking enforcement machinery.
- Inadequacy infrastructure and supporting services
- Affordability and playability and
- Socio-economic and socio-cultural determinant.

An intelligent management strategy has to take cognizance of key parameters such as hydrology, fish stocks and dynamics of their population together with regulatory measures for fishing. Observance of closed seasons and setting up of fish sanctuaries has proved their efficacy in the faster recovery of impaired fisheries.

Bibliography

A. Solaimalai, N. Ravisankar and B. Chandrasekar: *Farming Systems: Theory and Practice*, International Book Distributors, 2005.

A.S. Rawat: *Alternative Farming Systems in Dry Temperate Zone of Himachal Pradesh*, Indus, Delhi, 2002.

Ajit Kumar Roy and Niranjan Sarangi: *Applied Bioinformatics Statistics and Economics in Fisheries Research*, New India Publication Agency, Delhi, 2008.

Ajit Kumar Roy: *Evaluation and Impact Assessment of Technologies and Developmental Activities in Agriculture, Fisheries, and Allied Fields*, New India Publishing Agency, Delhi, 2011.

Amarjeet Singh: *Animal Husbandry*, Rajat Publication, Delhi, 2008.

Arnold C. Long: *Deep Sea Demersal Fish and Fisheries*, Cyber Tech Publication, Newe Delhi, 2009.

Aruna T. Kumar: *Handbook of Animal Husbandry*, Indian Council of Agricultural Research, 2008.

Arvind N Shukla: *Encyclopaedia of Fish and Fisheries, Vols. 1 to 5*, Discovery Publishing House, Delhi, 2008.

Ashok Kumar: *Animal Husbandry*, Discovery, Publication, Delhi, 2006.

B K Singh: *Applied Fisheries and Aquaculture*, Swastik Publication, Delhi, 2008.

B.L. Jana: *Farming Systems*, Agrotech Publication, Academy, 2014.

B.R. Selvamani and R.K. Mahadevan: *Fish Farming Systems*, Campus Books International, Delhi, 2008.

Gary T. Sakagawa: *Assessment Methodologies and Management: Proceedings of the World Fisheries Congress, Theme 5*, Oxford & IBH, Delhi, 1995.

Gyan Deep Singh: *Animal Husbandry*, Anmol, Publication, Delhi, 2008.

H.R. Singh and W.S. Lakra: *Coldwater Aquaculture and Fisheries*, Narendra Publication, Delhi, 2000.

Hrishikes Bhattacharya: *Commercial Exploitation of Fisheries : Production, Marketing, and Finance Strategies*, Oxford University Press, Delhi, 2002.

J.K. Roshan: *A-Z Fish and Fisheries*, Centrum Press, Delhi, 2009.

J.S. Datta Munshi, J. Ojha and T.K. Ghosh: *Advances in Fish Research: Vol: 3: Fisheries and Fish Biology Research*, Narendra Publication, Delhi, 2003.

K. C. Pandey: *Concepts of Indian Fisheries*, Shree Publishers , Delhi, 2012.

K.P. Biswas: *Ecological and Fisheries Development in Wetlands : A Study of Chilka Lagoon*, Daya Publication, Delhi, 1995.

K.P. Biswas: *Economics in Commercial Fisheries*, Daya Publication, Delhi, 2006.

Kamal Kishore Singh: *Fisheries and Aquatic Resources of India*, Sonali Publication, Delhi, 2011.

L. Lal: *Animal Husbandry (Objective Fundamentals)*, Agrotech, Publication, Delhi, 2010.

M. Youssouf Ali: *Fish, Water and People : Reflections on Inland Openwater Fisheries Resources of Bangladesh*, University Press, Delhi, 2001.

Manoj Kumar Rai: *Textbook of Animal Husbandry*, Oxford Book Company, Delhi, 2012.

Nishamani Kar: *Animal Husbandry and Rural Development*, Deep and Deep, Publication, Delhi, 2002.

Pratap Narain; M P Singh; Amal Kar; S Kathju and Praveen Kumar: *Diversification of Arid Farming Systems*, Scientific Publication, Delhi, 2008.

R Renaville and A Burny: *Biotechnology in Animal Husbandry*, Springer Publication, Delhi, 2008.

Rajendra Reddy and J.P. Abhay Shankar: *Animal Husbandry*, Commonwealth, Delhi, 2008.

Renaville: *Biotechnology in Animal Husbandry*, Springer, Publication, Delhi, 2004.

S.C. Panda: *Cropping and Farming Systems*, Agrobios, Publication, Delhi, 2004.

Shagufta Jamal and H P S Arya: *Participatory Rural Appraisal in Agriculture and Animal Husbandry*, Concept, Publication, Delhi, 2004.

U.K. Behera: *A Textbook of Farming Systems*, Agrotech Publishing Academy, 2014.

U.K. Behera: *Advances in Farming Systems*, Agrotech Publication, Delhi, 2012.

Vishal Anand: *Encyclopaedia of Fish and Fisheries*, Anmol Publication, Delhi, 2006.

Yasin Khan: *Animal Husbandry and Farming Management*, Agrotech Press, Delhi, 2013.

Yougesh Kumar and Rajeev Tyagi: *Aquaculture Fisheries Biotechnology and Genetics*, Manglam Publishers, Delhi, 2013.

Index